Translanguaging as Transformation

RESEARCHING MULTILINGUALLY

Series Editors: **Prue Holmes**, *Durham University, UK*, **Richard Fay**, *University of Manchester, UK* and **Jane Andrews**, *University of the West of England, UK*

Consulting Editor: Alison Phipps, *University of Glasgow, UK*

The increasingly diverse character of many societies means that many researchers may now find themselves engaging with multilingual opportunities and complexities as they design, carry out and disseminate their research. This may be the case regardless of whether or not there is an explicit language and multilingual aspect to their research. This book series proposes to address the methodological, practical, ethical and other options and dilemmas that researchers face as they go about their research. How do they design their research methodology to account for multilingual possibilities and practices? How do they manage such linguistic complexities in the research domain? What are the implications for their research outcomes? Research methods training programmes only rarely address these questions and there is, as yet, only a limited literature available. This series proposes to establish a new track of theoretical, methodological, and ethical researcher praxis that researchers can draw upon in research(er) contexts where multiple languages are at play or might be purposefully used. In particular, the series proposes to offer critical and interpretive perspectives on research practices and endeavours in inter- and multi-disciplinary contexts and especially where languages, and the people speaking and using them, are under pressure, pain, and tension.

All books in this series are externally peer-reviewed.

Full details of all the books in this series and of all our other publications can be found on http://www.multilingual-matters.com, or by writing to Multilingual Matters, St Nicholas House, 31-34 High Street, Bristol BS1 2AW, UK.

Translanguaging as Transformation

The Collaborative Construction of New Linguistic Realities

Edited by
Emilee Moore, Jessica Bradley and James Simpson

MULTILINGUAL MATTERS
Bristol • Blue Ridge Summit

DOI https://doi.org/10.21832/MOORE8045

Library of Congress Cataloging in Publication Data
A catalog record for this book is available from the Library of Congress.
Names: Moore, Emilee, editor. | Bradley, Jessica, MA editor. | Simpson,
    James, 1967- editor.
Title: Translanguaging as Transformation: The Collaborative Construction
    of New Linguistic Realities/Edited by Emilee Moore, Jessica Bradley,
    James Simpson.
Description: Bristol, UK ; Blue Ridge Summit : Multilingual Matters, 2020.
    | Series: Researching Multilingually: 3 | Includes bibliographical
    references and index. | Summary: "This book examines translanguaging as
    a resource which can disrupt the privileging of particular voices, and a
    social practice which enables collaboration within and across groups of
    people. The chapters critically examine how people work together to
    catalyse change in diverse global contexts" — Provided by publisher.
Identifiers: LCCN 2019054077 (print) | LCCN 2019054078 (ebook) | ISBN
    9781788928038 (paperback) | ISBN 9781788928045 (hardback) | ISBN
    9781788928052 (pdf) | ISBN 9781788928069 (epub) | ISBN 9781788928076
    (kindle edition) Subjects: LCSH: Translanguaging (Linguistics)
Classification: LCC P115.35 .T73 2020 (print) | LCC P115.35 (ebook) | DDC
    404/.2—dc23 LC record available at https://lccn.loc.gov/2019054077
LC ebook record available at https://lccn.loc.gov/2019054078

British Library Cataloguing in Publication Data
A catalogue entry for this book is available from the British Library.

ISBN-13: 978-1-78892-804-5 (hbk)
ISBN-13: 978-1-78892-803-8 (pbk)

Multilingual Matters
UK: St Nicholas House, 31-34 High Street, Bristol BS1 2AW, UK.
USA: NBN, Blue Ridge Summit, PA, USA.

Website: www.multilingual-matters.com
Twitter: Multi_Ling_Mat
Facebook: https://www.facebook.com/multilingualmatters
Blog: www.channelviewpublications.wordpress.com

The policy of Multilingual Matters/Channel View Publications is to use papers
that are natural, renewable and recyclable products, made from wood grown in
sustainable forests. In the manufacturing process of our books, and to further
support our policy, preference is given to printers that have FSC and PEFC
Chain of Custody certification. The FSC and/or PEFC logos will appear on those
books where full certification has been granted to the printer concerned.

Typeset by Riverside Publishing Solutions.

Contents

Contributors

Joëlle Aden is full Professor in the School of Social Sciences and Humanities at the University of Paris-Est Créteil and works in the area of language education in connection with the performing arts. She is a faculty member of the Paris-Créteil Graduate School of Teacher Education and a full member of the IMAGER Research Group within which she leads a research team. In her latest projects she looks at how a performative approach to language teaching through the lens of the enaction paradigm (Varela) might pave the way for new perspectives in a plurilingual and pluricultural context.

Jane Andrews teaches and researches in a range of areas within the field of education and has a particular interest in multilingualism and learning. Jane supervises doctoral students and jointly leads the Professional Doctorate in Education (EdD) programme. Current research interests include children's perspectives on being multilingual, researching multilingually and using creative arts in teaching children developing English as an Additional Language in schools.

Louise Atkinson holds a PhD in Fine Art from the University of Leeds and is a freelance visual artist, curator and researcher. Her work explores ideas of co-production and cross-cultural understanding in and through material culture, as a way of understanding how these contribute to notions of place. Through involving participants in her artistic and research practice, she considers how individual voices and experiences are represented in changing constructions of heritage. Her work is accessible on her website: www.louiseatkinson.co.uk.

Camilo Ballena is a Wichi language teacher at the Centre for Research and Training for the Aboriginal Modality (CIFMA) and at the Universidad del Chaco Austral (UNCAUS). He is a member of the *Consejo de la Lengua Wichí* (Wichi Language Board) and an external researcher for the Centre for Language Studies in Society (CELES). He authors a guide for adult literacy in Wichi and several works on Wichi language teaching. Currently, he conducts research on the transmission and use of Wichi writing and leads the Wichi Lhomet publishing house, devoted to the publication of didactic materials in Wichi.

Mike Baynham is Emeritus Professor of TESOL at the University of Leeds and a fellow of the Academy of Social Sciences. He is a sociolinguist by training and an applied linguist by affiliation. His research interests include literacy studies, language and migration, particularly migration narratives, multilingualism and translation. He was a co-investigator on the AHRC funded TLANG (Translation and Translanguaging) project and his monograph *Translation and Translanguaging*, with T.K. Lee, is published by Routledge (Baynham & Lee, 2019). He has also edited, with Luiz Paulo Moita Lopes, a theme issue of *AILA Review* entitled 'Meaning Making in the Periphery' (Moita Lopes & Baynham, 2017).

Martha Bigelow is Professor in Second Language Education at the University of Minnesota, USA. Her research focuses on equity in language teaching and learning with particular emphasis on contexts with immigrant and refugee-background youth. She has engaged in transformative and sustainable curriculum projects in Vietnam, Costa Rica and India. Her book with Doris Warriner is entitled *Critical Reflections on Research Methods: Power and Equity in Complex Multilingual Contexts* (Multilingual Matters).

Adrian Blackledge is Professor of Sociolinguistics in the Faculty of Social Sciences at University of Stirling. His publications include *The Routledge Handbook of Language and Superdiversity* (with Angela Creese, 2018), *Heteroglossia as Practice and Pedagogy* (with Angela Creese, 2014), *The Routledge Handbook of Multilingualism* (with Marilyn Martin-Jones and Angela Creese, 2012), *Multilingualism, A Critical Perspective* (with Angela Creese, 2010). He was Co-Investigator for the Arts and Humanities Research Council Translating Cultures Large Grant: 'Translation and Translanguaging: Investigating Linguistic and Cultural Transformations in Superdiverse Wards in Four UK Cities'.

Jessica Bradley is Lecturer in Literacies in the School of Education, University of Sheffield, UK. Her doctoral research 'Translation and Translanguaging in Production and Performance in Community Arts' (2018) considered translanguaging and text trajectories in street arts production and performance. She is interested in young people's understandings of language and culture through linguistic landscapes, collaborative ethnography and creative practice. She co-directs the Literacies Research Cluster at the University of Sheffield and co-convenes the AILA Research Network on Creative Inquiry in Applied Linguistics.

Angela Creese is Professor of Linguistic Ethnography in the Faculty of Social Sciences at the University of Stirling. Her disciplinary home is interpretive sociolinguistics and she draws on theories and methodologies from linguistic anthropology to investigate language in social life. Angela has led several large Research Council Grants (AHRC and ESRC) on multilingualism in city and school contexts and has been advancing ideas on heteroglossia, translanguaging and superdiversity as ideological orientations to social and linguistic diversity. Her research draws on empirical data gained through ethnographic observations, audio and video recordings of interviews and everyday interactions to which she brings an ethnographically informed discourse analytic approach. Her publications include *The Routledge Handbook of Language and Superdiversity* (with Adrian Blackledge), *Linguistic Ethnography* (with Fiona Copland), *Heteroglossia as Practice and Pedagogy* (with Adrian Blackledge, 2014, Springer); *The Routledge Handbook of Multilingualism* (2012, with Marilyn Martin-Jones and Adrian Blackledge); *Multilingualism: A Critical Perspective* (with Adrian Blackledge, 2010, Continuum); Volume 9: *Ecology of Language, Encyclopedia of Language and Education* (2009); *Teacher Collaboration and Talk in Multilingual Classrooms* (2005) and *Multilingual Classroom Ecologies* (2003).

Sandrine Eschenauer is Associate Professor at the Aix-Marseille University (LPL UMR 7309) in the field of language sciences/didactics of foreign languages. She has experience of primary and secondary school levels, as well as university. She is particularly involved in 'arts and languages' educational programs. Her research focuses on the translanguaging aspects observed while using performance (in particular through theatre) to teach two foreign languages at the same time (dual language classes), within a multicultural environment. Through the arts, she encourages teachers and learners to unearth their creative potential, a process that echoes the biological roots of learning as a holistic phenomenon which is both individual and inter-reactive.

Richard Fay is Senior Lecturer (TESOL and Intercultural Communication) at The University of Manchester. He teaches and researches critical applied linguistics areas including: appropriate methodology and intercultural communication in TESOL; researching multilingually, the languaging of research, and the risk of a linguistic contribution to epistemic injustice; intercultural ethics for knowledge-work in an era of increasing interconnectivity; ecological perspectives on researcher praxis and researcher education, and the case for developing translingual researcher mindset; and narratives of linguistically-shaped identity. He also has interests in world music ensemble pedagogy.

Katja Frimberger, PhD, is an independent researcher and educator interested in socio-aesthetic learning processes and how we arrive at formulations of 'value' in participatory art-making. Together with filmmaker Simon Bishopp, Katja has run two participatory arts projects: a Creative-Scotland funded film education initiative for young people with refugee experience (http://showmanmedia.co.uk/scotland/) and the Paul Hamlyn-funded 'U-Animate', using the latest performance-capture technology to involve care-experienced children and young people in animation and storytelling.

Ofelia García is Professor in the PhD programs of Urban Education and of Latin American, Iberian, and Latino Cultures (LAILAC) at the Graduate Center of the City University of New York. She has been Professor of Bilingual Education at Columbia University's Teachers College, Dean of the School of Education at the Brooklyn Campus of Long Island University, and Professor of Education at The City College of New York. Among her best-known books are *Bilingual Education in the 21st Century: A Global Perspective*; *Translanguaging: Language, Bilingualism and Education* (with Li Wei, 2015, British Association for Applied Linguistics Book Award recipient). Her books include *The Oxford Handbook of Language and Society* (with N. Flores & M. Spotti); *Encyclopedia of Bilingual and Multilingual Education* (with A. Lin & S. May), *The Translanguaging Classroom* (with S. I. Johnson & K. Seltzer); *Translanguaging with Multilingual Students* (with T. Kleyn). García was co-principal investigator of CUNY-NYSIEB (www.cuny-nysieb.org) from its inception in 2011 until 2016. García's extensive publication record on bilingualism and the education of bilinguals is grounded in her life experience living in New York City after leaving Cuba at the age of 11, teaching language minority students bilingually, educating bilingual and ESL teachers, and working with doctoral students researching these topics.

Lou Harvey grew up in a musical family who encouraged her love of singing, stories and drama, and after studying English Language and Literature she taught English as a Foreign Language in Slovakia and the UK. This professional background combines with her twin hobbies of choral singing and fiction writing to inform her research. Based in the Centre for Inclusion, Childhood and Youth at University of Leeds, she specialises in learning at the intersection of language and the arts, particularly in intercultural, public engagement and post-conflict contexts. Her current interests lie in theorising the relationship between learning and voice in arts-based, collaborative and co-produced research. She co-convenes the AILA Research Network on Creative Inquiry in Applied Linguistics.

Margaret R. Hawkins is a Professor in the Department of Curriculum and Instruction and the PhD program in Second Language Acquisition at the University of Wisconsin-Madison. Her work, centered on engaged scholarship around issues of equity and social justice, focuses on languages, literacies and learning in classroom, home, and community-based settings in domestic and global contexts. Her published work examines: classroom ecologies; families and schools; language teacher education; global digital partnerships for youth; and responses of new destination communities to mobile populations. She has worked with schools, communities, community-based organizations and institutions of higher education locally, nationally and globally.

Kendall A. King is a Professor of Second Language Education at the University of Minnesota. Her scholarship examines ideological, interactional and policy perspectives on second language learning and bilingualism, with particular attention to educational practices impacting language use among Indigenous populations in Latin America and Spanish and Somali speakers in the US. She teaches graduate-level courses in sociolinguistics, language policy, language research methods, and language education and undergraduate courses in linguistics, and is incoming president of the American Association of Applied Linguistics

Lotta Kokkonen works as a Lecturer at the Centre for Multilingual Academic Communication, University of Jyväskylä. Her areas of teaching include speech communication and intercultural communication. She defended her PhD on refugees' interpersonal relationships in 2010, and her research interests include asylum seekers' and refugees' social networks and belonging, international students' wellbeing and social networks, and networking from a relational perspective. She has been involved in curricula development for students' exchange periods and is responsible for organizing the study programme for exchange students. She has also coordinated a research project on highly-educated immigrants' language learning and belonging.

Maija Lappalainen works as a coordinator at the Centre for Applied Language Studies at the University of Jyväskylä. She has been working in assisting roles on projects related to adult migrant language education, labour migration and integration into work communities and everyday life and struggle in asylum-seeking settings.

Li Wei is Chair of Applied Linguistics and Director of the UCL Centre of Applied Linguistics, UCL Institute of Education, University College

London, UK. He is a Fellow of the Academy of Social Sciences, UK, and the Founding Editor of *International Journal of Bilingualism* and *Applied Linguistics Review*. His 2014 book with Ofelia García, *Translanguaging: Language, Bilingualism and Education*, won the 2016 British Association for Applied Linguistics Book Prize.

Júlia Llompart-Esbert is a member of the Research Centre for Plurilingual Teaching and Interaction (GREIP). Her PhD analysed the socialisation processes and plurilingual practices of students of immigrant origin in Barcelona. She is a postdoctoral researcher at the Universitat Autònoma de Barcelona for the LISTiac project (*Linguistically Sensitive Teaching in all Classrooms*) and continues her research on plurilingual practices and education and collaborative and participatory action-research.

Dolors Masats works as a teacher-trainer and researcher at the Universitat Autònoma de Barcelona and as a consultant for the Ministry of Education and Culture of Andorra on the design of a project-based cross-disciplinary curriculum. As a member of the Research Centre for Plurilingual Teaching and Interaction (GREIP) she leads or participates in numerous classroom-based national and international research projects in the field of conversational analysis applied to language learning in multilingual and multicultural milieus. She is a founding member of EDiLIC (*Éducation et Diversité Linguistique et Culturelle*) International Association and co-authors various classroom materials to promote language awareness and plurilingual education.

Emilee Moore is a Serra Húnter Fellow (Assistant Professor) at the Universitat Autònoma de Barcelona. She is interested in language practices in multilingual and multicultural educational contexts from a perspective that integrates linguistic ethnography, interactional sociolinguistics, ethnomethodology and sociocultural learning theories. She helps develop primary and secondary school teachers who are prepared to educate children and youth in contexts of linguistic diversity. She is a member of the Research Centre for Plurilingual Teaching and Interaction (GREIP) at the Universitat Autònoma de Barcelona and co-convenor of the AILA Research Network on Creative Inquiry in Applied Linguistics.

Luci Nussbaum was Professor at the Universitat Autònoma de Barcelona until 2016 and leader of the Research Centre for Plurilingual Teaching and Interaction (GREIP) until 2014. She specialises in the study of plurilingual practices from interactional and sociolinguistic perspectives.

Sari Pöyhönen is Professor of Applied Linguistics and deputy head of department at the Centre for Applied Language Studies, University of Jyväskylä, Finland. Her research and writing (ca 100 publications, including over 30 peer-reviewed journal articles) focus on language, identity and belonging; minorities and language rights; migration, asylum and settlement; and adult migrant language education policies. Through linguistic ethnography, creative inquiry and narrative approaches she focuses on language issues within wider cultural and political contexts and social structures.

James Simpson lectures in Language Education at the School of Education, University of Leeds, UK. His research interests span multilingualism and language education, and include adult migrant language education practice and policy, and creative inquiry in applied linguistics. He is the co-author of *ESOL: A Critical Guide* (OUP, 2008, with Melanie Cooke), the editor of *The Routledge Handbook of Applied Linguistics* (2011), and the co-editor of three further books. He is the founder of the discussion forum, ESOL-Research and is active in migrant language education policy formation nationally and locally. He was a Co-Investigator on the AHRC-funded project 'Translation and translanguaging' (2014–2018).

Tawona Sitholé is a storyteller and musician, and co-founder of Seeds of Thought, a non-funded arts group. He is currently UNESCO artist-in-residence at the University of Glasgow, with research and teaching roles in the school of education and medical school. Better known as Ganyamatope (his ancestral family name) his heritage inspires him to make connections with other people through creativity, and the natural outlook to learn. He has widely published as a poet, playwright and short story author. Other educational roles are with Glasgow School of Art, University of the West of Scotland, University of Stirling and Newcastle University, and Scottish Book Trust.

Mirja Tarnanen is Professor of Language Education (Finnish as a L1 and L2 and literature) in the Department of Teacher Education at the University of Jyväskylä. Her research focuses on literacy and assessment practices across the curriculum, migrants in professional and educational communities, policies and practices in second language teaching, professional learning and agency of pre-service and in-service teachers.

Ginalda Tavares Manuel studied a BA in Spanish with Japanese at Manchester Metropolitan University. She is an alumna of Leeds Young Authors and has participated in and won poetry slams in Leeds and across the UK as part of that organisation.

Gameli Tordzro undertook his doctoral research as part of the AHRC-funded project 'Researching Multilingually at Borders of Language, The Body, Law and the State' and is Artist in Residence of the UNESCO Chair on Refugee Integration through Languages and the Arts (UNESCO RILA), based in the School of Education at the University of Glasgow. His academic and professional background is in theatre, traditional African music and film directing. He set up Ha Orchestra, the first symphonic African Orchestra in Scotland, in June 2014 as part of the cultural festival of the Glasgow 2014 Commonwealth Games. He has been artistic director of Pan African Arts Scotland since 2006.

Virginia Unamuno is an independent CONICET researcher at the Centre for Language Studies in Society (CELES) of the Universidad Nacional de San Martín (UNSAM), where she also teaches linguistics. As a sociolinguist, her research interests revolve around the study of language policies from ethnographic and interactional perspectives. She directs a project about the new uses and meanings in the transmission of indigenous languages in northern Argentina. She has written numerous academic works, among which her books *Lengua, Escuela y Diversidad Sociocultural* (Barcelona, Graó) and *Lenguaje y Educación* stand out. She is often a guest lecturer in various Argentine universities.

Claudia Vallejo Rubinstein is a PhD candidate, adjunct lecturer and member of the Research Centre for Plurilingual Teaching and Interaction (GREIP) at the Universitat Autònoma de Barcelona, where she has participated in local and international projects on plurilingualism and social inequalities in education. Her PhD research analyses plurilingual and pluricultural practices in an after-school literacy program for 'at risk' students.

Zhu Hua is Professor of Applied Linguistics and Communication in Birkbeck College, University of London. Among her publications are *Exploring Intercultural Communication: Language in Action* (2019, Routledge, 2nd edn), *Crossing Boundaries and Weaving Intercultural Work, Life, and Scholarship in Globalizing Universities* (2016, Routledge, with Adam Komisarof) and *Research Methods in Intercultural Communication* (2016, Blackwell). She is book series editor for Routledge Studies in Language and Intercultural Communication and Cambridge Key Topics in Applied Linguistics (with Claire Kramsch).

Acknowledgements

Our sense of gratitude on the publication of this volume is extensive, for its writing and editing has itself been a cooperative and collaborative process. First and foremost, we wish to thank each and every one of the contributors for their participation, and for their entirely positive and enthusiastic stance towards the process of the production of the volume.

We would also like to express our thanks to the team involved with the research from which this book project grew – the AHRC-funded project 'Translation and Translanguaging: Investigating Linguistic and Cultural Transformations in Superdiverse Wards in Four UK Cities' (TLANG). Special thanks are due to the team members who contributed directly to our book: Angela Creese, the Principal Investigator of TLANG, Mike Baynham, Adrian Blackledge, Li Wei and Zhu Hua. Thanks are also due to other members of the team, in particular Leeds-based researchers Jolana Hanusova and John Callaghan, our Key Participants, and our other collaborators, without whose engagement, enthusiasm and intellectual curiosity this book would not exist.

Thanks too to Ofelia García, not only for writing a thoughtful, insightful foreword, but also for inspiring us and so many others in the development of concepts around translanguaging.

We are grateful to all those who generously devoted their time and expertise to read, comment upon and otherwise help develop chapters in the volume, and in particular to the anonymous reviewer.

Our work together began at the School of Education, University of Leeds, and we would like to thank our colleagues there for providing a productive environment. The Research Centre for Plurilingual Teaching and Interaction (GREIP) in the Faculty of Education at the Universitat Autònoma de Barcelona has also offered support and inspiration both to Emilee and to several of the chapter authors. Jessica would like to thank colleagues in the School of Education at the University of Sheffield as well as the Educational Engagement team at the University of Leeds and Multilingual Manchester.

Anna Roderick, Laura Longworth and Flo McClelland at Multilingual Matters have been unstinting in their support, and we thank them most sincerely.

Finally, we owe the greatest debts of thanks to our families.

Foreword: Co-labor and Re-Performances

Ofelia García

Reading the chapters in this book reminds me of how knowledge is always becoming. Every time we engage in a process of *co-labor*, as Ballena, Masats and Unamuno call it in their contribution to the book, we also re-perform our experiences and understandings, as we broaden the perspective from which we had earlier been seeing and acting. For me, this volume confirms some of my understandings about the potential of translanguaging as transformative action, while extending and challenging others. Moore, Bradley and Simpson, as well as their collaborators, re-inscribe their meanings, as they ground their work in a political imagination that is committed to social transformation, and in a semiotic approach to repertoire which focuses on the performative.

This book adds to the growing literature on translanguaging, but it does so by starting from co-labor with communities, students and teachers. The translanguaging work here is performed across communities. Sometimes these are far apart, as in the globally-spread communities in Hawkins' chapter on critical cosmopolitanisms. Other times they are quite local and indigenous, such as the Wichi in El Chaco, Argentina, in the chapter by Ballena and colleagues. And yet other times, multilingual diverse students and their teachers perform within the same ESOL learning context, as in the chapter by Simpson. But the co-labor is not always bounded by community. In Parts II and III of the book, it happens across academic disciplines and ways of enacting meanings, including the arts, poetry, performance and service-learning, along with what we deem to be the linguistic, including literacy. Throughout the book, researchers, teachers and students co-labor together, as each of these concepts is transformed.

The approach taken in this book fits well with the theory of *translangageance* which Aden and Eschenauer develop in their chapter, a welcomed addition to the literature. Making reference to the work of Francisco Varela, translangageance focuses on *langage* (in the French sense), not simply as the linguistic, but also as corporal (the body in its movement) and cultural histories, called forth through the performances of our embodied minds in a process that is always emerging and that is interwoven with aesthetic experience. As such, becoming bilingual is not simply about adding to a language system, but is about reconfiguring the entire system through a social coupling in relation to self, to others and to the environment. The intra- and inter-actions shape each other, and so discursive practices are not merely representations but are truly matters of practices.

This book moves translanguaging in a promising direction precisely because it starts with the idea that language is not simply the linguistic, but refers to the social coupling that makes it possible for the body and mind to act together, to integrate actions, so that we exist as human beings in language. That is, unlike much of the work in translanguaging, including my own, the authors in this book do not simply go beyond language(s), or beyond the linguistic to encompass multimodalities, or beyond traditional language education pedagogies or traditional language education research. Instead the work here starts from the other end. The *beyond* is here naturalized because it does not emerge from static components but from the dynamic act of bringing everything together in a constantly emerging process. Thus, there is no discussion here, for example, of whether bilingual speakers' translanguaging encompasses a duality (in that named languages have their own separate grammars) as in the work of MacSwan (2017), or whether the repertoire is unitary as is claimed in the work of Otheguy *et al.* (2015, 2018). And there is no discussion here of whether named languages are being taken up as having been socially constructed and only *socially* real (as in Makoni & Pennycook, 2007, & much of my own work) or whether named languages also have psycholinguistic reality (as in those who continue to interpret translanguaging as code-switching). Given that the emphasis in this book is on the enactive-performative nature of the integrative process of meaning making, there is no need to enter into a discussion of the 'linguistic' within this approach.

Likewise, there is no need in the work in this book to discuss multimodalities as separate from the linguistic, or to argue about what belongs to linguistics and what to pragmatics. The separation is not important because again, the emphasis in all the chapters is on the integrative performative that takes up what the feminist economic geographers Gibson-Graham (2008) call the 'performative ontological project'. This project recognizes that our discourse becomes part of our existence.

Many of the chapters in this book examine how the social coupling that is language is performed among students and teachers in different types of educational settings – bilingual education programs, ESOL programs, local and transnational classrooms, community arts programs, after-school programs, service-learning projects and other undertakings. But what stands out in this pedagogical practice is the emphasis on collaboration, especially arts-based collaboration. The stance in this book focuses on how all components that have been seen and studied as separate (mind, body, the linguistic, the cultural, the emotional, the kinesthetic, the cognitive, the aesthetic, pragmatics, etc.) act together to make meaning that is ethical and capable of social transformations. Likewise, students and educators act together in this enactive-performative pedagogical practice. The understandings produced are inseparable from their bodies, language, cultural histories and aesthetic experiences.

The editors also remind us that the research in this book has been co-produced collaboratively. Researchers committed to social transformation cannot determine what communities want in terms of knowledge, understandings, language and literacy experiences. This must be done in and with communities. What distinguishes this book, then, is that the research agendas of the chapter authors have emerged from work done with and in communities, and with a commitment to social transformation. For example, the high school students in Barcelona in the chapter by Llompart-Esbert and Nussbaum are the researchers of their own communicative practices. And the collaboration between a poet (Tavares Manuel) and a sociolinguist (Moore) results in a chapter where the poetic/aesthetic and the academic come together as the work is co-read and co-written. The collaborative constructions that are taken up make room not only for community engagement and partnerships, but for true co-labor in the voicing of injustices and the bringing forth of different forms of activism.

For me, translanguaging has always contained within it the seeds of transformation – transformation that can only come about by disrupting the naturalizations concerning language and language education that have kept minoritized communities disengaged and miseducated. Language education has always served as a way to support processes of minoritization, racialization and the perdurance of coloniality. Translanguaging is not solely a scaffold to learn the dominant ways of using language; and it is not solely a pedagogy for those who are least able to succeed. Translanguaging is a way to enable language-minoritized communities who have been marginalized in schools and society to finally see (and hear) themselves as they are, as bilinguals who have a right to their own language practices, free of judgement from the white monolingual listening subject; and free to use their own practices to expand understandings.

So this book disrupts the position that the linguistic holds in our imaginations of language and language education as superordinate. In so doing, the concept of translanguaging is extended. It is not about simply moving beyond our naturalized categories of language, bilingualism and language education as some of us have been signaling (García & Li, 2014). In this book it is about a co-labor, a co-labor that focuses on the collaborative actions of people as they come together in blending speaker / hearer / student / teacher / researcher / poet / artist / author / writer / reader.

I must add two caveats to the more transcendental view of language performances taken up in this book. One is that if the social transformative aspect of translanguaging is to be preserved, the performative has to go beyond the aesthetic, bringing in the criticality that is required. That is, creativity, as Li Wei (2011) has said, has to go hand in hand with criticality. Beyond the simple notion that meaning emerges in action from moving bodies/the embodied mind, bodies are also positioned differently in society. So the question is not only how to integrate actions, but also to question why it is that certain bodies – black, brown, refugee, immigrant, First Nations, women, poor, queer, transgender – are positioned in ways that do not allow them to act as white heterosexual monolingual listening subjects deem legitimate. The aesthetic has an important role in opening up a translanguaging space where these bodies gain legitimate action, but the aesthetic cannot simply be reduced to an emotional reaction. Language-minoritized communities' pain, grief, hope and aesthetic/emotional knowledge can only be transformative if they result in challenges to the structural inequalities that would position these bodies differently from how they are positioned today.

The second caveat has to do with the contexts in which the enactive-performances take place. Many of the pedagogical contexts included in this book are experimental programs, after-school programs, innovative projects. How can mainstream schools be transformed in ways that allow the potential of an enactive-performative pedagogy to be fulfilled? Given that schools are instruments of nation-states precisely to control the heterogeneity and diversity in their midst (despite discursive attempts to embrace them), many of these innovative pedagogical efforts fall short. And yet, I deeply believe that the modernist structural view of language supported in schools, and of language education simply for travel and intercultural communication, need to be disrupted. Only by showing the potential of 'otherwise' pedagogical practices can a space be carved out of the monolithic practices and curricula that we presently find in schools.

Now that many authors have shown that translanguaging as a pedagogical practice can be used to liberate the meaning-making potential of marginalized students, and that translanguaging disrupts our modernist and structuralist understandings of language itself, it is time to ponder how structures that continue to produce social inequalities can be transformed. How do we permanently destroy the walls/muros that keep bodies positioned differently, with the powerful in order, and with disorder, chaos and marginality created among the others?

Audre Lorde, a New Yorker whose parents were from Jamaica, famously said in 1979 '[T]he master's tools will never dismantle the master's house. They may allow us temporarily to beat him at his own game, but they will never enable us to bring about genuine change' (1979, np). What are the tools then that must be used to dismantle the institutions of the masters? How do poor, black/brown, women, lesbians, indigenous, minoritized beings make their knowledges visible? Political trajectories of resistance cannot just lead to the knowledge of the masters, but must raise the visibility of other knowledge to equal status in order to develop what the Portuguese scholar Boaventura de Souza Santos (2007) has termed an *inter*knowledge. Unlike interculturality, this interknowledge performed through what we might call translanguaging, is not about building peaceful intercultural relationships (or having dual or multi-language competencies). In unleashing what Anzaldúa (1987) calls the wild tongue, language minoritized speakers perform life *entre mundos*/ in the borderlands (Anzaldúa, 1987), thus disrupting the equally false imaginaries of life on one side of the wall (the normal one, the dominant one, the white one) vs. the other side (the marginalized one, the wild one, the racialized one). It is the energy produced by performing *entre mundos* translanguaging that had been rendered as illegitimate, non-standard, incomplete, a result of interference, that may open up cracks in the solid muros that institutions like schools have created.

References

Anzaldúa, G. (1987) *Borderlands/La Frontera: The new mestiza*. San Francisco: Aunt Lute Books.

García, O. and Li, W. (2014) *Translanguaging: Language, Bilingualism and Education*. London: Palgrave Macmillan Pivot.

Gibson-Graham, J.K. (2008) Diverse economies: Performative practices for 'other worlds'. *Progress in Human Geography* 32 (5), 613–632.

Li, W. (2011) Moment analysis and translanguaging space: Discursive construction of identities by multilingual Chinese youth in Britain. *Journal of Pragmatics* 43, 1222–1235.

Lorde, A. (1979) The Master's Tools will Never Dismantle the Master's House. *History is a Weapon*. www.historyisaweapon.com/defcon1/lordedismantle.html (accessed January 2020)

MacSwan, J. (2017) A multilingual perspective on translanguaging. *American Educational Research Journal* 54 (1), 167–201.

Makoni, S. and Pennycook, A. (2007) *Disinventing and Reconstituting Languages*. Clevedon: Multilingual Matters.

Otheguy, R., García, O. and Reid, W. (2015) Clarifying translanguaging and deconstructing named languages: A perspective from linguistics. *Applied Linguistics Review* 6 (3), 281–307.

Otheguy, R., García, O. and Reid, W. (2018) A translanguaging view of the linguistic system of bilinguals. *Applied Linguistics Review* 10 (4), 625–651

Santos, B. de S. (2007) Beyond abyssal thinking: From global lines to ecologies of knowledges. *Review (Fernand Braudel Center)* 30 (1), 45–89.

Translanguaging as Transformation: The Collaborative Construction of New Linguistic Realities

Jessica Bradley, Emilee Moore and James Simpson

Translanguaging, in Li Wei's words, has 'captured people's imagination' (2017: 9). While the number of academic articles and books about translanguaging has grown exponentially since the early 2000s, the title of this edited volume includes two further key terms – *collaboration* and *transformation* – which encapsulate the specific contribution we offer the field. The authors of the 12 chapters describe how collaboration and transformation are integral to their endeavours across different geographical locations and spaces of engagement. The aim of this collection, therefore, is not solely to describe diverse practices of translanguaging, or places and ways in which translanguaging might be enabled, but also to critically examine how people work together to catalyse change. Such change relates to how people and their communicative resources are positioned in different localities, and equally importantly, how people with different backgrounds, different frames of knowledge and different needs come together, communicate and work together. Such *intra-action* (Barad, 2007) – the co-constitution of entangled agencies from within encounters between people, the material world and discourses – offers the possibility of including different viewpoints for imagining alternative linguistic realities, and thus new methodologies for understanding and constructing worlds *for* and *through* language. Li conceptualises translanguaging as a practical theory of language, in that it involves an ongoing and emergent '*process of knowledge construction* that goes beyond language(s)' (2017: 15). In contributing to this practical theory, our volume presents translanguaging as being both an

1

ontological and an epistemological project: an endeavour concerned not only with the meaning of language and communicative practices, but also with how such meanings are generated.

The three parts of the book articulate diverse voices speaking of and from different experiences and traditions in the study of language and linguistic diversity; the authors also use words other than translanguaging to refer (in similar albeit nuanced ways) to the communicative practices which they examine. These include *translangageance* (Aden & Eschenauer), *transmodalities* (Hawkins), *bilingualism* (Ballena *et al.*), *multilingualism* (Andrews *et al.*, Simpson), *plurilingualism* (Llompart-Esbert & Nussbaum), *pluriliteracies* (Vallejo Rubinstein) and *voice* (Harvey). Indeed, compared with some of these terms, translanguaging is a relatively new concept that can be used for describing practices that are not themselves inherently novel. It is also one that has been assumed differently in the local contexts and scholarly traditions represented in the book. García, for example, makes explicit reference to aspects such as race, social class and sexuality in her understanding of how translanguaging works, while these aspects are less explicit in at least some other contributions to the volume. Translanguaging has emerged as part of a broader, critical process in which the meaning of 'language', understandings of 'language practices', and ideas about how knowledge of these phenomena is generated have themselves been objects of transformation. The traditional understanding of 'language(s)' as monolithic construct(s) existing independently of communicative use has been rejected in fields including interactional sociolinguistics, linguistic ethnography and critical applied linguistics in favour of conceptualisations of *languaging* (Becker, 1995) as practical social action that draws on an expansive repertoire of (not only linguistic) semiotic resources (Gumperz, 1964; Lüdi & Py, 2009, Blommaert & Backus, 2011; Rymes, 2014). Like the authors in this volume, many researchers now consider the concept of translanguaging beyond language(s) and encompass within their focus the multimodal nature of communication (Blackledge & Creese, 2017; Bradley & Moore, 2018; Kusters *et al.*, 2017; Zhu Hua *et al.*, 2017). Translanguaging thus reflects the multiplicity, fluidity, mobility, locality and globality of the resources deployed by individuals for engaging in complex meaning-making processes. It provides 'a way of capturing the expanded complex practices of speakers who could not avoid having had languages inscribed in their body' (García & Li, 2014: 18).

We do not dwell on whether the notions and terminology employed by authors in this volume are more or less appropriate for referring to the communicative contexts and encounters that they study (see, for example, Jaspers & Madsen, 2016; MacSwan, 2017; and Pennycook, 2016 for critical discussions in this area). Rather, we regard all the contributions as being important in generating emergent under-standings of language and other semiotic practices in contexts of diversity. This is what García is referring to, in the foreword to this volume, when she claims that 'there is no need to enter into a discussion of the "linguistic" within this approach'. By extension, as García describes, the chapters illuminate the relationships, processes and outcomes of collaborative endeavour (which blurs the boundaries between role), therefore broadening the translanguaging scope to encompass these practices.

The paradigm opened up by translanguaging (and similar co-existing notions) for the study of communication in contexts of diversity has allowed for a plethora of hidden and perhaps stigmatised ways of communication, often engaged in by members of linguistic minority groups, to be brought to the forefront of theory. In the words of feminist economic geographers Gibson-Graham, research and action have made the everyday practices of diverse individuals '"real", more credible, more viable as objects of policy and activism, more present as everyday realities that touch our lives and dynamically shape our futures' (Gibson-Graham, 2008: 618). This 'performative ontological project' involves seeing knowledge as always in a process of being and becoming, and scholars as privileged actors in this process of (re)inscribing meanings onto the world. Yet the *re*performance of reality, Gibson-Graham claim, requires 'new' academic subjects with an orientation towards a 'new' ethical practice. Thus, while much research has confirmed the dominance of certain communicative spaces and practices and the oppression of others, ethical practice would invite researchers to open spaces of freedom and possibility, by de-exoticising supposedly omnipresent forms of power in a way that new realities may be imagined and constructed. A performative ontological project is therefore intrinsic to collaborative, co-produced and action research agendas. It is also closely related to a transformative activist stance, described by Stetsenko as a way of researching 'that transcends the separation between theory and practice while embracing human agency grounded in political imagination and commitment to social transformation' (Stetsenko, 2015: 102).

The possibility of re-imagining and re-constructing linguistic realities – as a performative ontological project – is where trans-languaging research, we argue, holds most promise. In their extended definition of translanguaging, García and Li (2014) encourage scholars to engage in research that is trans-system, trans-space, trans-disciplinary and *transformative*, in seeking to go between and beyond socially constructed spaces, systems and practices of knowledge production in enacting novel ways of engaging with language, cognition, social relations, education, and social structures. From an epistemological perspective, translanguaging thus also offers new means of understanding knowledge production and of engaging as researchers with communities and their members in novel and mutually beneficial ways. With the aim of contributing to the epistemological as well as methodological turn that a translanguaging approach could potentially afford research into language and linguistic diversity, our volume showcases studies that embed long-term community partnerships in their processes, with a strong social justice orientation. Indeed, by referring to *new* in the title of the volume we are not only alluding to the fluidity and mobility of the social meanings attached to language uses. Perhaps even more importantly we are hinting at the possibility of modifying subjectivities through the types of action-oriented collaborations presented in the chapters.

The 'Translation and Translanguaging' Project

The inspiration for this book was the collective involvement of the editors and a number of the contributors in the UK Arts and Humanities Research Council-funded project 'Translation and Translanguaging: Investigating Linguistic and Cultural Trans-formations in Superdiverse Wards in Four UK Cities' (henceforth TLANG[1]). The three linking chapters are written by scholars involved in the TLANG project as co-investigators, Mike Baynham, Adrian Blackledge, Zhu Hua and Li Wei, and the projects' principal investigator, Angela Creese, has contributed an afterword. TLANG was a four-year project (2014–2018) in which collaboration between academic researchers, non-academic partners and community stakeholders was fundamental. Also central to the project was the concept of understanding, in this case understanding how people communicate across diverse languages and cultures in superdiverse cities (Vertovec, 2007). Translanguaging for the TLANG project was initially conceived as a means by which people make use of the communicative resources available to them in and across multiple inner-city and online spaces and places. These locations included food markets, libraries, advocacy charities, enterprise meetings and sports clubs, as well as the home. Founded upon a linguistic ethnographic approach, the research extended to encompass the multimodal, the embodied and what Baynham and colleagues describe as the 'trans-discursive' (Baynham *et al.*, 2015). Considering this mosaic of everyday spaces of everyday practice enabled new understandings of how people communicate to get things done: how they build their lives. In this way, translanguaging, as configured in the TLANG project and within the chapters in this volume, is hopeful practice. The 'trans-' approaches embodied across the TLANG project illustrate a postmodern orientation towards research which accepts and foregrounds an understanding of the 'complexities of the ethnic and linguistic mingling which takes place in social spaces worldwide' (Holliday & MacDonald, 2019: 10), spaces that are always emergent. The TLANG project therefore served as a catalyst for creative approaches to doing research and to understanding everyday communication in social life. At a methodological level it opened up a space for further development of lines of thinking and lines of doing. It formed, in this sense, a point of departure – and not least for this book.

Underpinning Themes

We now return to the foundational themes for our volume – *collaboration* and *transformation* – setting out some of the different ways in which these concepts frame the contributions.

Collaboration

How does our volume as a whole understand the concept of collaboration? The contributions are organised into three parts, focusing in turn on collaborative *relationships*, *processes* and *outcomes*. All three parts offer perspectives on collaboration, or on how diverse people, with different ways of knowing and doing, take action together. Or as Creese puts it in her afterword, the volume explores 'ways to open up the research process so that other voices can be heard in research accounts'. Part 1 focuses on the relationships that are built for and through research, Part 2 foregrounds the processes of joint engagement, and Part 3 centres on the outcomes of such shared work. As is the case with translanguaging, the contributors also use other terms to refer to their collaborative efforts, including *participatory research* (Llompart-Esbert & Nussbaum), *interthinking* (Andrews *et al.*), *co-labor* (Ballena *et al.*), *boundary crossing* (Moore & Tavares Manuel), *bricolage* (Bradley & Atkinson), *coproduction* (Pöyhönen *et al.*), *transauthorship* and *transcreation* (Harvey). The volume addresses the gap in accounts of what collaboration might mean in research, or the challenge of 'developing a language to talk about the different traditions that constitute the field' (Facer & Pahl, 2017: 4). Through their different contributions, all authors show how this 'language' for talking about collaboration is necessarily locally contingent and multi-voiced.

Part 1 of the volume, *Collaborative Relationships*, focuses on the broad range of contexts in which research is built and takes place. There is a risk in laying these upon the page. Often relational processes go unwritten, with research write-ups focusing on results, analysis and outputs. Here we showcase some of the messiness involved in the performative ontological project of translanguaging – the long-term relationships required, the commitments to being uncomfortable and to allowing different stories and understandings to come to the fore. The contributors do not seek to offer a guidebook to developing collaborative relationships for research. Instead they foreground the multiple ways in which this development can take place, from deliberate collaborations to emergent co-productions.

Part 2, *Collaborative Processes*, emphasises the negotiated and unfolding nature of collaborative work. The four chapters all focus on creative arts, and on collaborations between university researchers and creative practitioners, demonstrating how flexible and emergent epistemologies work in practice. The chapters show how collaborative processes challenge the ownership of knowledges in terms of whose voices are heard and whose voices are therefore considered worth hearing. Moreover, these glimpses of the lived realities of doing research *with* people embody opportunities for creative ethnographic practice.

The chapters in Part 3, *Collaborative Outcomes*, demonstrate that the products of collaborative processes are inextricable from these processes and from the relationships established. The outcomes here perform the 'trans-', allowing insights into the potential transformative possibilities, and also to the potential restrictions. The reflections offered within the chapters build on the question of 'whose voices?' to ask 'whose *transformation*?'

In this way, the volume speaks to questions about the valuing of collaborative research, both inside and outside the academy, an issue raised by Facer and Pahl (2017). These authors state that in much research of this kind, there is a tendency to focus on the complexities of setting up a 'partnership' – and that work is needed to explore how research in this trans-space can be judged. We do not offer a taxonomy or a structured framework here. We do however seek to show glimpses, albeit partial ones, of the often invisible sides of research and make a link between the performative ontological project and the messiness inherent in the approaches taken by the chapter authors.

Transformation

In terms of how the contributions conceptualise transformation, the chapters across the volume typically focus on one of two types of change – either on practical manifestations of changes in people's practices in local settings, or on changes in subjectivities as a result of the relationships developed by participants in collaborative engagements. Jaspers (2018) presents a critique of translanguaging scholarship, problematising its transformative claims. Although his main focus is on research which describes the transformative potential of translanguaging pedagogies in schools, his words remind us to act with caution in upholding transformation as a causal effect of translanguaging, or even as its purpose. Indeed, one contribution to this volume (King & Bigelow) explicitly questions the chances of long-term transformations of communicative practices and pedagogies that are partially determined by fluctuating socio-political environments. In fact, all chapter authors present rigorous accounts not only of the changes their work hoped to bring about, but also of its limitations, and the challenges they encountered. One of our intentions as editors was to critically engage with notions of transformation and its (im) possibilities, paying attention to different aspects of research practice and allowing for further questions to emerge in the place of claims to exacting findings. By foregrounding collaboration across all of the contributions, the work also helps address questions such as 'who gets to decide what transformation is?', and 'how might transformation be experienced, in different localities?' We see transformation as a subjective experience, and the chapters help illuminate some of those subjectivities.

Outline

We now turn to the structure of the book and some brief notes on the content of the different contributions. Each of its three parts contains four main chapters, and is introduced by a short comment chapter.

In his comment chapter framing Part 1, *Collaborative Relationships*, Baynham notes that the relationships that are imagined and enacted by those engaged in researching translanguaging are of many kinds: relationships of solidarity (based on the transgression of linguistic and communicative boundaries) and of challenge (subverting prevailing normativities), as well as research relationships with participants. These have the aim of amplifying voice, enabling their 'speaking back to ideologies and norms of monolingualism and separate bilingualism in a way that might be politically effective' (Baynham, this volume). This 'freedom-orientation' of translanguaging is clearly evident in the main chapters, which highlight the centrality of relationships and dialogue within research contexts. Hawkins' chapter details the complexities of engaging youth from across the world in a film-sharing project, where she elaborates the potential of transmodal transnational interaction for generating critical cosmopolitanism with her participants. Translanguaging possibilities in adult migrant language education in the UK are the focus of Simpson's chapter: this is a domain of educational practice where normative understandings of monolingualism dominate, and where the concept of translanguaging is often at odds with established understandings of language use and language pedagogy. A different kind of tension is found in the context of a collaborative photography project with unaccompanied minors who are residents at an asylum seeker reception centre in the rural west of Finland, in the chapter by Pöyhönen and colleagues. Here a light is shone on how the sustained engagement of ethnography enables the establishment of relationships between the different actors involved in the project. Part 1 ends with a study of *co-labor* in the context of plurilingual education in Wichi-speaking communities in Argentina, by Ballena *et al*. These colleagues understand co-labor as working together with indigenous communities, and from symmetrical positions of power.

Authors of chapters in Part 2, *Collaborative Processes*, relate their work to the experience of carrying out collaborative research. The dialogic nature of the chapters emphasises, in the most positive way, how – as Blackledge notes in his framing comment chapter – once dialogism is recognised as a mode of representation, the authority of the single voice is questioned. Creative practice that is collaborative, as Blackledge explains using the example of poetry, enables the productive expansion and elaboration of interpretations of social life. Part 2 begins with Aden and Eschenauer's work in language education. They describe an enactive language pedagogy which they develop within a framework of what they term *translangageance*, focusing on the process of emergence of a 'common language'. Andrews and colleagues in the next chapter reflect on their use of arts-based methods in a large multi-site trans-national cross-disciplinary collaborative project. They regard their practice as transformative and consider this from the perspectives of the new materialism and from *interthinking*, which they gloss as collaborative problem-solving in teams across communicative modes. The chapter that follows also considers collaborative practice in the arts, this time at the intersection of research, practice and engagement. Here, Bradley and Atkinson explain how they use *bricolage* as a conceptual framework for their transdisciplinary pedagogical approach to the study of the linguistic and semiotic landscape (and its limitations), reflecting on the transformational affordances of translanguaging, in work with young people in Leeds, UK. In the last chapter in Part 2, a poet (Tavares) and an academic (Moore), describe their experience of creating co-produced research, also in Leeds, with a Youth Spoken Word poetry organisation. In so doing, they illuminate the processes involved in co-reading, co-interpreting and co-writing ethnography.

The *Collaborative Outcomes* of relationships and processes are the focus of chapters in Part III. The authors of the framing comment, Zhu and Li, caution that the outcomes described in these chapters cannot be divorced from process. Rather, one should consider outcomes themselves as 'a collaborative, creative and critical process, underpinned by strong commitment to social justice and equality'

(Zhu & Li, this volume). Harvey, in the first chapter of Part 3, discusses the Bakhtinian concept of *voice*, the foundation of her work with a theatre company in the north of England in the co-production of performance art. She develops the concept in relation to this work, describing voice as the material vehicle for processes of trans-ing: translanguaging, transcreation, and transauthorship. In the chapter that follows, by King and Bigelow, we return to migrant language education contexts, but this time in Minnesota, US. The work they discuss, on translanguaging pedagogies, takes place with adolescents who are refugees and are in the process of becoming literate as well as being language learners. Plurilingual learning in a High School in Barcelona is examined in the next chapter by Llompart-Esbert and Nussbaum: they demonstrate how the adolescent students who participate in their collaborative research are highly competent in using their linguistic and communicative repertoires for learning. The final chapter of the section, by Vallejo Rubenstein, also describes research in Barcelona. This time though the focus is on translanguaging as transformative pedagogy, in work that 'bridges' across educational contexts: an after-school literacy program in a multicultural, multilingual primary school, and a teacher education project where educational resources incorporating a translanguaging approach are developed for – and with – the participants in the after-school literacy program.

Looking Forward

The authors of the framing chapters, together with García and Creese in their foreword and afterword, already highlight many of the main ideas from the book to be taken forward in future translanguaging scholarship. In this final section we will briefly refer to some of the concrete findings or main conclusions that weave across chapters. Firstly, the volume offers important methodological insights for researchers and research institutions: it illuminates the affordances of collaborative research for generating 'newness' as it is perceived in each local context by different individuals. As highlighted by Facer and Pahl (2017), critical work is still needed in terms of how this research and its 'impact' is valued both inside and outside the academy. The volume offers clear methodological advances in terms of interdisciplinary research with creative practitioners, as pointed out by Creese in her afterword. Some of the chapters are co-authored with artists (Andrews *et al.*, Bradley & Atkinson, Moore & Tavares Manuel), and these as well as others (Aden & Eschenauer, Harvey; Pöyhönen *et al.*) report on the development of creative inquiry methodologies for answering language-related research questions. In Applied Linguistics there is increasing interest in creativity and the arts (broadly understood) as method for researching language and as communicative practice (see the AILA Research Network on Creative Inquiry in Applied Linguistics, formed in 2018; Bradley & Harvey, 2019), and this volume offers methodological advances for this emerging field. García and Creese in their forward and afterword argue that further interrogation of the relationship between criticality and creativity in translanguaging research and practice emerges from the volume as a necessary future direction for this type of work. While not addressing creative inquiry specifically, other chapters offer insightful examples of how translanguaging research may bridge spaces (for example, a non-formal educational context and a university teacher education course in Vallejo Rubinstein's study) and roles (for example, the dual role of teacher and researcher taken on by Llompart-Esbert in her study with Nussbaum), and others are co-written by academics and educational activists (for example, Ballena *et al.*), all advancing our understandings of what collaborative research means.

As a number of the chapters focus on formal or non-formal educational contexts, the volume also offers significant findings in terms of translanguaging pedagogies. One important conclusion in this regard is the recognition that letting 'other' languages into teaching/learning spaces does not necessarily mean that educators and learners need to use 'all languages at all times'. Indeed, in her own work in the field of education, García speaks of translanguaging as 'part of the discursive regimes that students in the 21st century must perform, part of the linguistic repertoire that includes, at times, the ability to function in the standardized academic languages required in schools' (García & Sylvan, 2011: 389). The recognition that standard languaging practices in 'one language at a time' are also part of translanguaging emerges in the chapter by Ballena *et al.*, who support spaces for learning Wichi, spaces for learning Spanish as well as bilingual spaces. It is evident too in the chapters by Llompart-Esbert and Nussbaum and by Hawkins, in which learners' full communicative repertoires are included in learning processes which progress towards and culminate in project products (digital stories, digital posters) that use one language only, alongside other semiotic modes. Explicit recognition of translanguaging as a continuum of language practices could help reduce the sort of tensions reported on in the chapter by Simpson in terms of pervasive monolingual ideologies in educational settings. A further important finding in terms of translanguaging pedagogies is clearly represented by Vallejo Rubinstein's chapter, in which educational resources were developed 'ground up' from the practices of plurilingual students. Translanguaging pedagogies are necessarily responsive to local realities, and thus may look different from one context to another.

Finally, the chapters point to tensions in terms of the need for 'fixity', for example in placing translanguaging in durable policies. In their chapter, King and Bigelow speak precisely of the experimental nature of many of the policies and practices supporting translanguaging. Simpson discusses the complexities of engaging with policy actors, and sees most promise in collaborative grass-roots initiatives. Similarly, Ballena *et al.* conclude that enduring educational transformations 'can only occur through [...] sustained, reflective and collective actions', rather than through particular policies or action plans. These findings align with our understandings of transformation and newness, discussed in earlier sections of this introductory chapter, as being intricately tied to collaborative endeavour and individual subjectivities.

Note

(1) See: www.tlang.org.uk

References

Barad, K. (2007) *Meeting the Universe Halfway: Quantum Physics and the Entanglement of Matter and Meaning.* Durham: Duke University Press.

Baynham, M., Bradley, J., Callaghan, J., Hanusova, J. and Simpson, J. (2015) Translanguaging business: Unpredictability and precarity in superdiverse inner city Leeds. *Working Papers in Translanguaging and Translation* 4. https://tlang754703143 .files.wordpress.com/2018/08/translanguaging-business.pdf (accessed January 2020)

Becker, A. (1995) *Beyond Translation: Essays towards a modern philology.* Ann Arbor, MI: University of Michigan Press.

Blackledge, A. and Creese, A. (2017) Translanguaging and the body. *International Journal of Multilingualism* 14 (3), 250–268.

Blommaert, J. and Backus, A. (2011) Repertoires revisited: 'Knowing language' in superdiversity. *Working Papers in Urban Language and Literacies* 67. www.kcl. ac.uk/ecs/research/research-centres/ldc/publications/workingpapers/abstracts/ wp067-repertoires-revisited-knowing-language-in-superdiversity (accessed January 2020)

Bradley, J. and Harvey, L. (2019) Creative inquiry in applied linguistics: Researching language and communication through visual and performing arts. In C. Wright, L. Harvey and J. Simpson (eds) *Voices and Practices in Applied Linguistics: Diversifying a Discipline* (Chapter 6). York: White Rose University Press. https:// universitypress.whiterose.ac.uk/site/books/10.22599/BAAL1/

Bradley, J. and Moore, E. (2018) Resemiotisation and creative production: Extending the translanguaging lens. In A. Sherris and E. Adami (eds) *Making Signs, Translanguaging Ethnographies: Exploring Urban, Rural and Educational Spaces* (pp. 81–101). Bristol: Multilingual Matters.

Facer, K. and Pahl, K. (2017) *Valuing Interdisciplinary Collaborative Research: Beyond Impact.* Bristol: Bristol University Press.

García, O. and Li, W. (2014) *Translanguaging: Language, Bilingualism and Education.* Basingstoke: Palgrave Macmillan.

García, O. and Sylvan, E. (2011) Pedagogies and practices in multilingual classrooms: Singularities in pluralities. *The Modern Language Journal* 95 (3), 385–400.

Gibson-Graham, J.K. (2008) Diverse economies: Performative practices for other worlds. *Progress in Human Geography* 32 (5), 613–632.

Gumperz, J.J. (1964) Linguistic and social interaction in two communities. *American Anthropologist* 6 (2), 137–53.

Holliday, A. and MacDonald, M.N. (2019) Researching the intercultural: Intersubjectivity and the problem with postpositivism. *Applied Linguistics*. https://doi.org/10.1093/applin/amz006 (accessed January 2020)

Jaspers, J. (2018) The transformative limits of translanguaging. *Language and Communication* 58, 1–10.

Jaspers, J. and Madsen, L.M. (2016) Sociolinguistics in a languagised world: Introduction. *Applied Linguistics Review* 7 (3), 235–258.

Kusters, A., Spotti, R., Swanwick, R. and Tapio, E. (2017) Beyond languages, beyond modalities: Transforming the study of semiotic repertoires. *International Journal of Multilingualism* 14 (3), 219–232.

Li, W. (2017) Translanguaging as a practical theory of language. *Applied Linguistics* 39 (1), 9–30.

Lüdi, G. and Py, B. (2009) To be or not to be … a plurilingual speaker. *International Journal of Multilingualism* 6 (2), 154–167.

MacSwan, J. (2017) A multilingual perspective on translanguaging. *American Educational Research Journal* 54 (1), 167–201.

Pennycook, A.D. (2016) Mobile times, mobile terms: The trans-super-poly-metro movement. In N. Coupland (ed.) *Sociolinguistics: Theoretical Debates* (pp. 201–217). Cambridge: Cambridge University Press.

Rymes, B. (2014) *Communicating Beyond Language*. New York: Routledge.

Stetsenko, A. (2015) Theory for and as social practice of realizing the future implications from a transformative activist stance. In J. Martin, J. Sugarman and K.L. Slaney (eds) *The Wiley Handbook of Theoretical and Philosophical Psychology* (pp. 102–116). Malden, MA: John Wiley and Sons.

Vertovec, S. (2007) Super-diversity and its implications. *Ethnic and Racial Studies* 30 (6), 1024–1054.

Zhu, H., Li, W. and Lyons, A. (2017) Polish shop(ping) as translanguaging space. *Journal of Social Semiotics* 27 (4), 411–433.

Part 1: Collaborative Relationships

Comment on Part 1: Collaborative Relationships

Mike Baynham

The study of translanguaging has moved away from its initial focus on multilingual resources in the repertoire to consider more broadly the interaction of other semiotic materialities, bringing together the verbal/visual/embodied. In this process it has become clearer that translanguaging is transgressive of the boundaries set by monolingual or separate bilingual ideologies. Moreover, in this more developed formulation, translanguaging aligns with the theoretical move which challenges the privileging of language in relation to other semiotic orders. Translanguaging, as 'multilingualism from below', to adapt the phrase of Pennycook and Otsuji (2015), is no respecter of boundaries and as such, as pointed out in Baynham and Hanušová (2018), is liable to attract censure for boundary transgression. The chapters brought together in this section provide ample evidence of this. So while on one hand we can welcome the celebratory emphasis on 'a free and active subject who has amassed a repertoire of resources and who activates this repertoire according to his/her need, knowledge or whims, modifying or combining them where necessary' (Lüdi & Py, 2009: 159), we have to acknowledge on the other hand that the freedom and ability to be active in this way are resources that are not evenly distributed in a globalised world whose terrains are marked by sharp inequalities of all sorts, including linguistic (see García, this volume). In other words, that freedom has to be struggled for, not assumed. So if translanguaging is multilingualism from below, with connotations of Spivak's (1988) subaltern, then it can be seen as a creative/political project *vis à vis* the normativities of monolingualism and separate bilingualism, in other words a *speaking back*.

Of course speaking back means nothing if nobody hears. To have a voice and not be heard is to experience pain. So a necessary response for speaking to become action is *audibility*, being heard. It is perhaps no accident that this move to celebrate the creativity of multilingualism from below has become associated not just with academic argument but also with art practice of various sorts: public statements/events/ interventions which can literally serve to amplify the dimension of challenge brought about by translanguaging as speaking back to monolingual and separate bilingual normativities. So this creative practice, this celebration, is also a political struggle against these normativities, again something that is concretely demonstrated in contributions to this section. One way of approaching the idea of speaking back and audibility is through Bakhtin's notion of dialogic and authoritative speech (Bakhtin, 1981). Authoritative speech for Bakhtin is speech that expects no answer, while dialogic speech is speech that expects and invites an answer, that answer provoking a further response and so on. Within this frame, speaking back to dominant monolingual/separate bilingual normativities is to force the authoritative into dialogue, to insist on being heard. This is why we need to recognise that as a project of creativity oriented to freedom, as suggested by Lüdi and Py and now by many others, it is also a political project which both imagines and enacts other kinds of relationship.

The last point brings us to the theme of this section, *Collaborative Relationships*. We have the relations suggested by translanguaging itself as a phenomenon identified as multilingualism from below, having on the one hand its relationships of solidarity based on the transgression or flouting of boundaries and on the other hand relationships of challenge, undercutting and subverting prevailing normativities. We also have the relationships which researchers might seek to establish in researching translanguaging as multilingualism from below, desiring both to create and maintain dialogic relations with research 'subjects' and to amplify their speaking back to ideologies and norms of monolingualism and separate bilingualism in a way that might be politically effective. Again the papers in this section provide plenty of concrete evidence of such processes. So the question has to be: how does the researcher manage to enable dialogic and co-productive research relations with subjects, in the face of manifest inequalities of power/knowledge/resources and carry out research that can be understood, in terms of the framing of this volume, as *transformative?*

To engage with these issues I will review each chapter in turn, starting with Margaret R. Hawkins on critical cosmopolitanism in a transnational, transmodal video exchange project (Global Story-Bridges), exemplified through discussion of an exchange of videos between young people in a working class suburb of Barcelona and a village in Uganda. Hawkins' framework highlights some of the points I have been making above: that despite much glib talk in sociolinguistics over the years about globalisation and mobility, the full enjoyment of their benefits in the neoliberal world order have been typically the domain of elites; that there are tensions, as Stuart Hall (2006 and as cited by Hawkins) points out, between the erasure of difference and the hostile exaggeration of difference. Disturbingly the world seems to be moving into a phase of the hostile exaggeration of difference and inhospitality and the cruelty of borders, which makes the project of critical cosmopolitanism that Hawkins advocates politically crucial to counter a drawing back behind national borders. And finally that the struggle for critical cosmopolitanism has to contend with the inequalities along every dimension imaginable which riddle what Wallerstein (1974) calls the world system.

Hawkins' project aims to create online spaces to grow such potential for critical cosmopolitanism among school age young people through exchanging videos made to illustrate their daily lives. Yet the data she discusses show the difficulties of sharing across lifeworlds, supporting the idea that critical cosmopolitanism is hard, indeed painful, work. The Barcelona-produced video, filmed in an informal afterschool club – that is a leisure context – projects play and leisure, attracting, it seems, a certain amount of mild censure from the Ugandan participants for whom karaoke seems to be a lurid night club activity. The attention of the Ugandan youths in contrast is drawn to a garlic planting activity, which resonates with their own rural agricultural context. They are curious about the type of fertilizer used. In turn, the typical afterschool activities shown in the Ugandan video involve domestic chores in the house and garden, including drawing and carrying water in jerry cans. A video response from Barcelona involves playful attempts to reproduce balancing a water container on the head, finishing the video by lightheartedly and water-wastefully throwing the jugs in the air to see if they will land bottom up, provoking curiosity and confusion from the Ugandan youth. Hawkins concludes by making clear that the Ugandan youths found the Barcelona comments ('It's funny and it's a trend') hurtful ('For us this is real. This is our life.'). So the project of critical cosmopolitanism emerges as hard work and pain, working against the grain of differences in frame, inequalities and power relations, involving both reflection on the lives of others and also self-reflection.

James Simpson's contribution addresses the tensions inherent in challenging monolingual and separate bilingual norms in an educational context, drawing on data that surfaced during an online discussion on the ESOL Research Network mailing list. Insisting on English only in the classroom, he points out, is a long-established and often unexamined norm in English language teaching, a manifestation of that monolingual ideology which I discussed above. Posing questions in the online discussion concerning the role of multilingualism leads to the articulation of different positions corresponding to the monolingual norm and its challenge from a multilingual perspective.

This provides a concrete example of the point made above about the authoritative voice, here the classroom voice stating the rule 'English only in class!' which can be challenged and brought into dialogue with other perspectives using a device such as an online debate. There are clear differences in the perspectives articulated with some contributors arguing for the English only approach, while others advocate for bringing other languages into the classroom. This echoes the kind of differences that emerged in the Global StoryBridges project, in type if not in degree, showing that even in a specific subset of the English Language Teaching profession such differences can thrive, become embedded and need a dialogic approach to unpack the differing ideas and make them explicit.

It is worth noting here a recurrent gap between what speakers say they do and what they actually do. For example, in studies of code-switching in India carried out by Gumperz (e.g. 1964), participants in the study would often express strong disapproval of code-switching as a practice, then a few moments later resort to it in conversation. Similarly in bi/multilingual classrooms there may be a strongly enforced rule against bringing other languages into the classroom, which is routinely flouted in practice, as teachers try against the grain to create meaningful connections with what is being learnt in their students' heads. Even in a classroom where the teacher is adhering to an English only approach, further investigation into the ecology of the classroom might reveal students annotating worksheets in their own language, surreptitiously consulting an online bilingual dictionary or Google Translate, or falling back on whispered conversations with a classmate. This would suggest what one might call a submerged or suppressed multilingual substratum in otherwise English only contexts. Of course these data also support the point made earlier that there are distinct monolingual and separate bilingualism norms and ideologies in play that can lead to phenomena such as translanguaging being suppressed and othered. This potentially normative context for purely celebratory treatments of translanguaging is important to bear in mind, the correlate of this point being that advocating trans-languaging as a strategy is a political project which is likely to draw down critique from those taking the English only perspective. The political nature of both critical cosmopolitanism and multilingual pedagogies becomes even more crucial at a period when nationalism and its populist base is resurgent and borders are strengthened and a British Prime Minister can hope to score points with the electorate by stating that 'A citizen of the world is a citizen of nowhere' (e.g. Cooke & Peutrell, 2019; Grant, 2016; May 2016). In this phrase the then Prime Minister, Theresa May, argues for a narrow nationalism as the grounds for citizenship, rather than a critical cosmopolitanism. To counter the tendency towards these purely celebratory treatments of bilingualism there is a need to ground decisions about multilingual language use in the classroom in critical language awareness work, which makes visible the ideological tensions around language choice and in particular the link between nationalist resurgence and the closing of borders with the kind of linguistic embordering proposed by ideologies of monolingualism and separate bilingualism. This kind of explicit critical work about language choice can be found in the contribution from Ballena *et al.* below

Discussing Hawkins' contribution I said that developing critical cosmopolitanism is both hard and painful work since it involves working across difference of many dimensions, including unequal distribution of power/knowledge/resources. Working across difference in this way has a constant potential for misunderstanding and schema mismatch, and as we saw in Hawkins' paper, being misunderstood (not seen, not heard) causes pain. This of course lies at the heart of ethical questions in research, that those involved in the research should not be hurt by it. This hurt can be material but it can also be psychic. Such ethical questions that arise when one is working across difference in power/knowledge/resources in situations of precarity is very much implicated in the contribution from Sari Pöyhönen *et al.*, which focuses on the question of work with highly vulnerable subjects, unaccompanied adolescent refugees. I use the term 'subject' advisedly here to emphasise research participants as the subjects of their story, not the objects of research, not as subjected to research. The semantics of the word is slippery here, since to be a subject can also imply subjection to the research of others, that is to be an object for others. To work effectively with such complex social and affective dynamics involves working across distance and power/knowledge/resource imbalances to create relationship and trust which can form the basis for participatory practice.

The contribution of Pöyhönen and colleagues describes the building of relationship and trust through a photography project conducted in a hostel for unaccompanied adolescent refugees in Finland. The hostel is characterised by a great deal of fluidity and uncertainty on every level as the young people endure the ups and downs of asylum claims, of residents gaining asylum or losing it, or moving on for a variety of reasons. There is an important influence on the nature of belonging and the creating of a sense of homeliness in strange, as it were, *unheimlich* circumstances. Underpinning the research is an investigation into the nature of belonging in such circumstances and how the sense of belonging can be practically as well as theoretically engendered. There seemed to be some advantage here in focusing on the visual as a way of documenting and expressing the lived experience of the youths and the chapter describes a growth of trust and intimacy over the life of the project. As I have suggested this working with/on difference is hard emotional work and the chapter discusses the impact not just on the youths but on Gustaf, the project coordinator, who has to take time off work due to the emotional stress of his engagement in this work.

The aim of the research is to develop a relationship of knowledge and artistic co-production around a theme with the youths and their counsellors, so that they become, as I suggest above, the subjects of their stories, not the objects of the research of others. Despite the perceived advantages of working with the visual, with photography as a medium both for documenting and expressing feelings, the research also brings in discrepant voices, as when one of the youths is heard to wonder why they are wandering in the woods taking photos rather than doing something useful like learning Finnish. What we are engaging with here is not a neat and tidy research paradigm, but one that is messy and open-ended, allowing in discrepant voices, but one in which, arguably, relationship and trust are foregrounded, or rather relationship and trust emerge/are constructed over time.

The contribution by Camilo Ballena *et al.*, concerning the implementation of schooling and curriculum in Wichi, an indigenous language of Argentina, raises a number of issues that we have been addressing here. The chapter starts by framing the history of the re-emergence of Wichi as a language of education as part of a struggle against the monolingual language ideology which constructed Spanish as the national language of Argentina, while simultaneously marginalising the rich variety of indigenous languages (the authors mention 38). This theme connects with earlier discussion about the pain of having a voice but not being heard. The chapter documents this struggle for recognition, leading to the training of Wichi teachers first as teacher auxiliaries, later as fully qualified teachers, and initiatives to establish Wichi as a language of education alongside Spanish. Having established the historical background, the chapter describes a project between Wichi teachers and activists and university-based researchers. The relationship between the Wichi teachers and the university researchers is characterised not by the term 'collaboration' which is felt to be implicated in the unequal power/knowledge relationship typical of much such research, but by the term '*co-labor*', foregrounding the research investigation as joint work on an equal footing.

A persistent theme in these chapters is indeed to explore ways of transforming the power/knowledge/resources imbalance in order to undertake work as co-production or co-labor. Ballena and colleagues describe consultations with Wichi speaking parents to establish what they want from the school in terms of Spanish and Wichi, which showed that the parents wanted their children to have access to Spanish as the language of the wider society, but also to Wichi, not least because the recruitment of Wichi teachers had led to career possibilities for teachers of Wichi. Once dialogue is initiated, as in the situation described by Simpson, different voices/positions are heard. The outcome was the differentiating of the curriculum into spaces that were designated Spanish-only, Wichi-only and bilingual. This resonates with the discussion on language choice in the classroom reported in Simpson's contribution. Perhaps it is best to think of this solution as a way-station rather than a final state of Wichi language education in the Argentinian curriculum. If uses of Wichi continue to be progressively embedded in what one might call the public repertoire, then other solutions in the future might be imaginable. The authors show what the emergence of Wichi into public space has gone through, from the recognition in law in 1987 of the value of Wichi and other indigenous languages in education, followed in 2010 by the legal recognition of the co-officiality of Spanish with indigenous languages the initial recruitment of Wichi speakers as teacher auxiliaries, then later, due to political pressure, as fully trained teachers. Elsewhere, Ballena and Unamuno (2017) have documented the shift-in-progress in Wichi literacy practices, from a language that is read, to one that is increasingly written by Wichi writers who are thereby finding a voice in a range of contexts including social media. To put it another way, these Wichi speakers are extending their repertoire by adding Wichi as a language through which they have a voice in writing, contributing to the realisation of Lüdi and Py's characterisation of the free and active subject with a repertoire of resources. What is clear though, from Ballena and colleagues' contribution, is that this state of freedom and agency does not come without struggle, speaking back and the establishment of such freedoms within the legal framework and is thus, as I have argued in this chapter, painstaking and at times painful work.

References

Bakhtin, M. (1981) *The Dialogic Imagination*. ed. M. Holquist, trans. C. Emerson and M. Holquist. Austin: University of Texas Press.

Ballena, C. and Unamuno, V. (2017) Challenge from the margins: New uses and meanings of written practices in Wichi. In L.P. Moita Lopes and M. Baynham (eds) *Meaning Making in the Periphery*. AILA Review 30, 120–143.

Baynham, M. and J. Hanušová (2017) On the relationality of centres, peripheries and interactional regimes: Translanguaging in a community interpreting event. In L.P. Moita Lopes and M. Baynham (eds) *Meaning Making in the Periphery*. AILA Review 30, 144–166.

Cooke, M. and Peutrell, R. (2019) *Brokering Britain, Educating Citizens: Exploring ESOL and Citizenship*. Bristol: Multilingual Matters.

Grant, M. (2016) 'Citizen of the world'? Think again: British citizenship after Brexit. *Democratic Audit*, blog, www.democraticaudit.com/2016/11/21/citizen-of-the-world-think-again-british-citizenship-after-brexit/ (accessed 24 April 2019).

Gumperz, J.J. (1964) Hindi-Punjabi code-switching in Delhi. In H. Hunt (ed.) *Proceedings of the Ninth International Congress of Linguistics* (pp. 1115–1124). The Hague: Mouton.

Hall, S. (2006) *Interview of Stuart Hall March 2016*, www.youtube.com/watch?v=fBfPtRaGZPM (accessed 24 April 2019).

Lüdi, G. and Py, B. (2009) To be or not to be a plurilingual speaker. *International Journal of Multilingualism* 6 (2), 154–167.

May, T. (2016) *But if you believe you're a citizen of the world, you're a citizen of nowhere*, speech to the Conservative Party Conference 2016, https://blogs.spectator.co.uk/2016/10/full-text-theresa-mays-conference-speech/ (accessed 10 November 2019).

Pennycook, A. and Otsuji, E. (2015) *Metrolingualism: Language in the City*. London: Routledge.

Spivak, G.C. (1988) Can the subaltern speak? Revised edition, from the 'History' chapter of Critique of Postcolonial Reason. In R. Morris (ed.) *Can the Subaltern Speak?: Reflections on the History of an Idea* (pp. 21–78). New York: Columbia University Press.

Wallerstein, I. (1974) *The Modern World-System I: Capitalist Agriculture and the Origins of the European World-Economy in the Sixteenth Century*. New York: Academic Press.

1 Toward Critical Cosmopolitanism: Transmodal Transnational Engagements of Youth

Margaret R. Hawkins

The 'Trans-' Turn in Language Studies

Current discourses across disciplines that consider language and communications are turning to 'trans-' perspectives to signal a shift in conceptualisations of language in use, and indeed of today's world. If 'trans-' is taken to signal crossing, then our world is rapidly becoming transglobal, transnational and translocal, with human, material and semiotic flows moving with increasing volume and rapidity across it.

For some time now linguists have recognised that languages shift and change over time, intertwining with one another and with other semiotic resources to convey messages and meanings (e.g. Canagarajah, 2013, 2018; Hawkins, 2018; Toohey, 2018). Forms of communication matter and demand study, as do the messages and meanings that are conveyed and the resultant understandings between participants in semiotic exchanges. Put a bit differently, languages (and other semiotic modes and resources) matter in that they are vehicles for carrying meaning, and those meanings shape relations between people in the world. My thesis for this chapter is that local and global human relations matter – perhaps especially in our current era of transnational distrust and hostility – and that research and theory ought to attend to both the modes and means through which humans jointly construct understandings and the resultant impact on the relationships they forge.

In this chapter I will take up critical cosmopolitanism (Hawkins, 2014, 2018) to articulate why relations, identities and civic engagements matter in an era of rapid globalisation, and the importance of sustaining caring, open and equity-oriented dispositions toward global others. I define what a theory of *transmodalities* may offer to understandings of language-in-use, especially across distance, encompassing the broad array of semiotic resources that are leveraged in communication. Finally, I will illustrate the heuristic power of the dual lenses of critical cosmopolitanism and transmodalities through analysis of data drawn from Global StoryBridges (GSB) – a project in which global youth engage transnationally in digitally-mediated encounters – to illuminate understandings of semiotic flows in transnational communication and of the relationships and understandings they engender.

Translanguaging, Translingualism and Multimodalities

The notion of translanguaging (or translingual practice in Canagarajah's 2013 version) is gaining traction in fields attending to language. Focusing on languaging practices, it refers to the ways in which interlocutors flexibly and fluidly leverage their linguistic resources in communication (García & Li, 2014). In earlier accounts, e.g. code-switching (Heller, 1988) and crossing (Rampton & Charalambous, 2012), scholars theorised the ways in which people move between two or more named languages, whereas current 'trans-' perspectives position learners as fluidly drawing on (and among) all of the linguistic resources at their command in situated communicative acts. They mix and mesh codes and signs creatively and spontaneously, although in concert with and responsive to their contexts and interlocutors (Canagarajah, 2013; Li, 2018).

Beyond a narrow focus on language, however, scholars are attending to the importance of non-linguistic resources in communication, positing theories of multimodalities. Much of the work in multimodalities (particularly that following a Systemic Functional Linguistics orientation) has focused on technologies in communication and learning. How, for example, when we look at text (broadly defined, including images, video and so on) do we theorise semiotic resources in communication? Beyond language, what must we attend to in order to understand processes of communication? The multimodalities literature points to modes – e.g. visuals, music, gesture, gaze, and so on – that carry meaning both in isolation and in tandem with other modes they are enmeshed with in any communicative act (Jewitt, 2017; Kress, 2017). In multimodality, 'modes' are understood as 'a socially shaped and culturally given resource for meaning making' (Kress, 2017: 60), or as symbols that are recognisable to all involved in a communicative exchange. I problematise this, and draw on the notion of *transmodalities* (Hawkins, 2018, to be discussed below), as within forces of globalisation, communication increasingly occurs between people in different geographical locations who often do not share language and cultural backgrounds, norms and sign systems.

Yet it is important that, as strangers in a global world connect and communicate, they form caring, open and equity-oriented relationships and views of one another. To this end I developed GSB, in which youth in locations across the globe connect, share their lives, and forge relations.

Global StoryBridges

In GSB, English-learning youth in diverse global locales connect through digital stories (short videos) of their lives and communities that they make and share on a dedicated project website, and ensuing chat discussions. There are currently approximately a dozen sites in nine countries, all located in under-resourced communities. There are two 'clusters' of sites: one for youth who are 11–12 years of age; and the other for youth aged 14–18. The project is designed based on tenets of sociocultural learning theory, in that all activities are engaged in collaboratively, and learning occurs through situated interactions and negotiations of meaning. Each site has an adult facilitator that supports the work of the youth participants, but does not direct it, or overtly 'teach'. In this respect it differs from social networking; the youth collaboratively decide the focus and content of their videos, the processes of video making and roles they will take, and engage in the editing process together (within their site). They then post the final version. When a site posts a video, youth in other sites watch together (per site), engage in discussion, and together formulate questions and comments to post. When questions and comments are posted, the video makers watch them together, discuss them, and collaboratively decide how to respond. In this way all meanings and understandings are co-constructed both within and across sites, instantiating a Vygotskian approach to mediated learning and (scaffolded) co-construction of meaning (Vygotsky, 1978; Hawkins, 2010).

The goals of the project are to foster: (1) the language and literacy development of participating youth (including but not limited to English); (2) the development of technological skills; and (3) an awareness of global others and of oneself as a global citizen, through fostering (equitable and open) global relations. It is important to note that the project is not part of formal school curriculum; rather participating youth meet in out-of-school time, often in community-based spaces.

The project exemplifies 21st-century learning and engagement in that youth – some of whom have never interacted with those outside of their village, community or culture – are communicating trans-modally with diverse global counterparts. In order to explore the ways in which meanings are co-constructed (within and across sites) and relationships are developed, I invoke the dual lenses of *critical cosmopolitanism* and *transmodalities* (Hawkins, 2014, 2018), first discussing each, then applying them to data from communication between secondary-school-aged youth in Spain and Uganda.

Critical Cosmopolitanism

While globalisation references the ever-increasing rapidity of movements of people and things across the world (Appadurai, 1996), cosmopolitanism indexes human encounters within forces of globalisation. Although there is a significant body of literature on cosmopolitanism in many disciplines, dating back to early Greek civilisation, there are conceptual differences in understandings and applications of the term. In my work I have identified strands from the literature that speak to current conditions and have implications for human relations and communications in an increasingly globalised world, and for the education of youth (Hawkins, 2014, 2018). These strands are: global flows of resources and knowledge (e.g. Appadurai, 1996); citizenship, loyalties, affiliations and identities (e.g. Nussbaum, 1997; Rizvi, 2008); and ethics and caring (Appiah, 2006).

Sociologist Stuart Hall asserts that there are twin, and dialectical, dangers in globalisation: on one end the danger is 'an overriding sameness and homogenisation,' and on the other it is 'warring differences' (Hall, 2006). In other words, as actions, ideologies and initiatives of people everywhere on the globe increasingly and rapidly cross national and international boundaries to affect others, the danger is that we all become indistinguishable from one another as differences are erased, or conversely that our differences reign paramount and lead to disputes and disagreements. Cosmopolitanism offers a way to think about globalisation and difference that leads neither to erasure of difference nor to conflict, but rather, in the words of philosopher and cultural theorist Kwame Appiah (2006), suggests an understanding of 'universality plus difference' (2006: 151) that enables us to engage with and value such difference. That is the hope of Global StoryBridges: that youth participants will indeed explore universalities and differences together with local and global others, thus coming to a complex, multifaceted and layered view of human lives, identities and relationships.

However, a danger of cosmopolitanism resides in its dual meanings. Historically, the term came to signify the 'revolt of the elites against the low culture of the masses' (Delanty, 2006: 26). In today's vernacular 'cosmopolitan' has come to refer to an elite status-cosmopolitans are those who have the means and opportunities to travel, move between places and cultures, sample various ethnic cuisines, artifacts of cultures, languages, and so on. However, those who must move between places and cultures due, for example, to economic or political hardship would not, in this sense of the word, be considered cosmopolitan. Yet it is these people – migrants, immigrants, refugees, those from language and cultural groups other than those of the mainstream in the communities in which they live, and who often live in poverty – who in their everyday lives embody cosmopolitanism, or, to quote Hall (2006), find themselves 'living in translation every day of their lives.'

This distinction in social status has been noted, and termed 'cosmopolitanism on the ground' (Hansen, 2010), 'rooted cosmopolitanism' (Appiah, 2005), 'elementary cosmopolitanism' (Kromidas, 2011) and 'everyday cosmopolitanism' (Hull *et al.*, 2010). It seems to me that there are inherent societal dangers, as 'elite' cosmopolitans and 'everyday' cosmopolitans have little contact and reside in different sociopolitical spheres, with no opportunity to re-allocate and equalise social positions, possibilities and resources. Cosmopolitanism scholars, by and large, have failed to take into account differences in power, status and privilege among various members of and groups in society, such as those taken up by critical (and critical race) theorists, post-colonialist scholars and others, thereby missing crucial aspects of human and social relations. I, therefore, have issued a call for critical cosmopolitanism, to highlight issues of power, positioning and equity as central in the forging of human relations (Hawkins, 2014, 2018). In GSB, although all participating youth live in under-resourced communities, there are significant differences in their environments and the resources to which they have access, as well as linguistic, cultural and ideological differences, and these all play a role in the understandings they construct of one another and ultimately of themselves. Yet the goal of GSB is to create and sustain open, equitable and caring relations between global youth, in the hope of overcoming the sorts of stereotypes, biases and hostilities that are becoming ever more frequent in encounters across difference in our world.

Transmodalities

Critical cosmopolitanism provides a conceptual framework for global human relations – one that keeps issues of status, positioning and equity front and center – and GSB provides a venue through which global youth connect with one another. The challenge is to understand how the communicative, or semiotic, tools available to participants, and the affordances through which they are mediated, form the meaning-making among project youth that shapes the relationships they forge. Youth make and share videos representing their lives and communities, they ask and answer questions predicated on the video (or sometimes not). However, just putting youth in contact does not ensure that they will develop dispositions of inquiry, openness and caring toward one another; in fact, it may serve to reify existing biases and stereotypes. So to accomplish these goals, how do we understand meanings-under-construction? How do we analyze GSB's effect and impact? What sorts of analytic and heuristic tools are available for exploring semiotic processes in this work? While scholars working in multimodalities have made some headway in identifying communicative modes, and thinking about modal 'assemblages' (Latour, 2005), as noted earlier there are unique challenges in transnational communications that have been under-theorised and under-explored. Thus I developed the *transmodalities* framework to account for critical components of semiosis in transnational (and other) encounters, which is comprised of five 'complexities' (Hawkins, 2018).

Complexity #1: Modes intertwined

While multimodality identifies modes – such as gesture, posture, image, sound – and explores affordances of each, it attends less to the entanglement of modes, or the ways in which they mutually shape one another in particular constellations of use, having the whole communicative ensemble ultimately amount to more than the sum of its constituent parts. In transmodal communications, how then do we understand the ways in which modal components fluidly and continuously flow, converge, diverge and shift, mutually impacting one another, and how these movements and configurations shape the messages represented and received, and the design and processes of meaning-making?

Complexity #2: Relationships between modes, language and material objects

While scholars acknowledge that modes other than language play a part in communication, nonetheless language is positioned as central, with other aspects playing, in effect, a support role. Yet there are theories that place primacy on objects – material things – as carrying their own semiotic load. In such theories (e.g. Barad, 2003; Canagarajah, 2018; Latour, 2005) people and things together constitute production and meaning, and things can carry meaning even in the absence of spoken or written language, or of human design and intent. In transmodal communications, how can we account for the semiotic load of modes, language and material objects, and even de-center language, as we analyze communicative processes and messages?

Complexity #3: Production/assemblage, reception and negotiation

In multimodality literature, there is a spotlight on communicative design. That is, scholars attend to the ways in which modes are used in assembling messages. They address multimodal design (Lemke, 2002), multimodal production (Flewitt, 2011), multimodal orchestration (Jewitt, 2017), multimodal improvisation (Flewitt, 2011) and so on. This focus on how people use multiple modes to design messages infers human intent – messages are purposefully designed to carry specific meanings. However, messages do not always communicate the intent of their creator; they are imbued with meaning in concert with their context, things (and objects) that comprise them and mediate them, etc. Further, messages move through time and space, and across different groups of people, and interpretations are mediated by many factors (including space and place, as discussed in Hawkins, 2014). Thus I have posited the arc of communication as central to a theory of transmodalities. It includes acts of production/assemblage, but also those of movement, reception and negotiation. In transmodal communications, how do we consider the full arc of communication – what understandings are constructed among interlocutors as multimodal messages are assembled, travel through time and space, and are received and negotiated?

Complexity #4: Context and culture

Context, in a theory of transmodalities, refers to: the multimodal assemblage (or message) itself, as it provides the context for the interaction of modes within it; the attributes of space and place within which it is designed, across which it travels, and that in which it is received, as aspects of space and place mediate what can be conveyed and understood; and the histories and trajectories of the message, and of the people and places implicit in its creation, interpretation and negotiation.

Culture, on the other hand, refers to the shared knowledge, practices and beliefs of communities or groups of people. In transnational communications, when messages move between people or groups of people who do not share cultural beliefs, signs and symbols, ideologies, ways of life and understanding/being in the world, communication becomes more complex, as taken-for-granted representations and meanings cannot predictably be received as the message creator/s intended. In fact, terms such as 'language', 'culture', and 'community' become destabilised in a globalised world, as people, objects, ideas, knowledge and ideologies travel and take root in new places, and messages travel among people with no prior contact or encounters. In transmodal communications, how can we explore the arc of communication between disparate people (and groups of people), and how can ethnography help us understand emic perspectives and meanings (Kress, 2011), if we wish to forge productive and meaningful human relations.

Complexity #5: Transnationalism and relations of power

And, finally, as critical theorists have pointed out for quite some time, there are no human communications that exist outside of relations of power. Issues of identity, status and positioning are always integral to interactions – they shape them and are shaped by them – and to the meanings that are subsequently (co)constructed. In transmodal communications, power, status, and positioning may play out in many ways. As has been pointed out, access itself is an issue. Although Global StoryBridges only operates in under-resourced communities, youth come with differential access to and experience with technology. Poverty itself displays differently among sites in the videos produced. Youth understand one another through a lens of socioeconomic difference, and that gets reflected in the chat discussions. In a world so stratified by race, class, language, gender and (dis)ability, youth cannot shed these lenses, or ideologies, in understandings they construct of others and others' lives, nor ultimately in their understandings of their own. Thus a critical lens and reflexivity are crucial for global encounters, and must be deliberately fostered.

Global StoryBridges: Constructing Transglobal Relations

To explore why both critical cosmopolitanism and transmodalities matter in transnational interactions and communications I offer data from a set of exchanges between two secondary-level GSB sites; one in a community organisation in a rural Ugandan village, the other in an urban community center in Spain. The youth participants in Uganda were all indigenous to the community; they shared the same tribal and clan history, language and culture, going back many generations. Further, the community itself was not diverse, and they had little contact with global others. Their facilitator, too, was from their community. They met in a 'community library' – a small, simple structure in a remote area. In Spain, the majority of the youth were born locally, although some had parents or grandparents who had experienced internal migration and there was one girl born in Western Africa. They lived in a low socioeconomic status (SES) area of the Barcelona Metropolitan Area, a relatively large Spanish city. GSB there was under the auspices of a community program affiliated with a large university, located in an urban community center; the facilitator was an undergraduate at the university. Here we see, underscoring a point made earlier, that 'poverty' is a fluid and relative term, and that place-based resources are never identical or equal.

Recall that the project is meant to connect youth across the globe in service of developing awareness and understanding of global others (and of themselves as global citizens), developing dispositions of openness, inquiry and caring toward one another, and forging equitable relations. Here I will share communicative exchanges among the participating youth, then show what these twin lenses (critical cosmopolitanism and transmodalities) can enable us to understand about the content and process of meanings being constructed.

Data here are verbatim excerpts drawn from the videos posted by both sites and ensuing chat discussions.

'Our Second Video'

After an initial exchange of introductory videos, the Barcelona youth posted a video entitled, 'Our Second Video'. Recall that the youth are told that they are to make videos that represent their lives and their communities. The first screen in the video is text, and says, 'Hello, in this video will show you three different activities. (1) karaoke activity (2) planting garlic activity (3) sketch activity We hope you like it ;)'. They chose to include four discrete segments in this video. The first two segments are (two different) groups of them sitting in a room at the community center, in big overstuffed cushions and chairs, facing the camera, although with all eyes focused below the camera lens. The room is painted blue, with charts and graphics on the wall. There is music playing (Maroon 5), and the youth are staring at something unseen, attempting to sing along to the song (not very successfully), and each time the refrain rolls around ('would you still love me the same?'), they do a set of hand and body gestures. There is much giggling among this group of 4, and they clearly are having fun. This continues for 1 minute and 55 seconds, then the second group does the same with a different song (7 Years, by Lukas Graham). Although there are 7 youth in this second group, what is represented is much the same. The youth sit, in casual clothing, in comfortable chairs and cushions, stare at something just offscreen, and attempt to sing along to the song. This segment continues for 2 minutes and 5 seconds. After it ends, a screen appears with printed words: 'Thanks for Watching and Special Kiss to Photo:)'.

The next segment is entitled, 'Garlic Project', and begins with one of the project youth sitting in a field digging garlic sprouts out of the dirt. While digging, the facilitator asks questions which the boy answers: 'What are you doing Alex?' 'I am collecting some soil' 'Why are you collecting some soil?' 'Garlics' 'Are you going to plant garlics?'

'Yes'. The scene then shifts to the community center, where youth are pouring dirt from a bag (which reads Campo Sana) into a container. The facilitator narrates, 'Now we are mixing the soil we get from the park with this compost'. The next minute or so shows a group of youth mixing the soil and compost in a plastic bag with their hands. The video then cuts to a table with four plastic containers, each filled with dirt and labeled. This shot is subtitled, reading: 'As the las [sic] step: we water the garlic', and a girl proceeds to do just that. Throughout all of the video Catalan and Spanish are used among the youth, and by voices in the background. Credits scroll, and we move to the last segment of the video.

This segment is a fiction piece that a subset of the youth penned, scripted and shot. It is titled, 'The Revange of the Coins', and shows each of the coins they use, attributing personalities and competitive natures to them. So, for example, they show a 10-Euro cent coin, and narrate, 'am the most little of the family'. Then a 20-Euro cent coin, saying, 'I am the most handsome'. And so on. They then film coins 'escaping' from a wallet, and exacting revenge from their 'enemies', narrating through the coins' voices (e.g. 'we jump into the hands of the enemies'). At the end, one coin ends up on top of a slide and jumps off, knocking a girl in the head. The entire video lasts for almost 7 minutes.

Youth in Uganda watched, and, as per project routines, had a (facilitated) discussion about it. Afterwards they together composed then posted comments and questions for the Barcelona makers. The Barcelona youth, in turn, read the comments and questions, and responded. Here I will represent a subset of these exchanges as questions followed by responses, although the questions and responses were posted as separate entries per site. Offerings from Uganda are identified by 'U', and those from Barcelona by 'B'.

(U)   What is Karaoke in your understanding?, because this way in Uganda we understand it differently.

(B)   karaoke is singing a song while you are reading the lyrics on the screen and you are listening to the music

(U)   Are you people watching while singing?, because its like all your eyes are focused on one point and the same song is playing in the back ground.

(B)   yes we are reading the lyrics

(U)   What is compo sana made of?

(B)   Peat, perlite (to facilitate drainage of plants) lime and organic fertilize

(U)   Which type of soil was that young man scooping in the valley?

(B)   we don't know, it was normal soil from the valley, our place here

(U)   How long does garlic take to mature?

(B)   six months

(U)   The containers for growing the garlic were labelled are they of different types?

(B)   yes they come for different recycled bottles

(U)   Were you people in the class room during the Karaoke activity?

(B)   no, we were in the youth center.

(U)   Did you people live out the sketch activity or its the same as the revenge for coin.

(B)   all the sketches are unreal. We created the ideas.

(U)   Is the growing of garlic one of your projects or you just did it for video purpose

(B)   is one of the projects of young center

Upon receipt of the Barcelona responses, the Uganda youth posted another set of comments and questions. They said:

Our friends here are some comments and questions on your responses.

Here in Uganda we take karaoke to that time in disco clubs where ladies come to dance half naked or in pants.

What is peat and perlite made of and what quantity is required to make up that packet of compo sana?

Is the garlic for consuption or commercial purposes?

In a brief analysis, we see deep inquiry and engagement between the two groups, and an avid curiosity about each other's lives and activities. In terms of the five complexities of transmodalities, the video intertwines modes (music, movement, gaze, gesture, text as subtitles, captions and scrolling text, transitions, pace, and much more) to offer a complex representation of Barcelona youth, their community and center, what they do, and who they are, in multifaceted and layered ways. The material objects are deeply implicated in meaning-making, as the room in the community center contains comfortable seats, equipment to watch and hear music, colorful walls and environmental print (something not encountered in this rural Uganda village, and clearly confusing as the Ugandan youth ask whether that is a school classroom). The garlic roots, and the park itself, as well as the indoor gardening in plastic jugs, connote a very different lifestyle and sense of crops and agriculture. The fiction sketch, including the personification of coins is novel (and equally confusing), which leads to the next two categories: the arc of communication and context and culture.

The youth makers designed their 'activities' to convey specific (and different) messages. The meanings were not clearly conveyed, however, in large part due to differences in context and culture. In this Ugandan community all families, including children, are deeply involved in agricultural practices, because what is grown is what is available to eat. In Barcelona, on the other hand, the youth live urban lives, and buy their food. This difference can be noted in the predominance of questions on the garlic segment of the video, and the specificity of the questions. Ugandan youth are not only experts in farming and gardening because of their lived experiences, but also because, correspondingly, crops and agriculture are a major part of Ugandan school curriculum. Thus their attention was drawn to the wording on the bag of compost, and they wanted to know exactly what it was. The Barcelona youth could read what it said on the bag, but did not have deeper knowledge. The Ugandan youth wanted to know what kind of soil; the Barcelona youth replied that it was 'normal soil'. This is an example of how what can be noticed and learned is mediated by place.

There are many other misunderstandings represented in the chat texts, but I will here only focus on misunderstandings around karaoke because, in this exchange, it shows the import of the arc of communication. Barcelona youth wanted to portray something they do for fun, which was incomprehensible to the Uganda youth. They did not know what their peers were doing, what they were looking at, or what the word 'karaoke' meant in the opening text. So they asked, and the response together with the video portrayal enabled an emergent understanding. They then were able to identify one source of miscommunication: different understandings of what 'karaoke' is (perhaps as interpreted through certain moral and religious lenses), illustrating that the act of production led to a particular reception, which was then negotiated so that both parties gained insight into the others' understandings and lives.

Rather than a more in-depth analysis of this video, I will describe the next, posted by Uganda, to follow the semiotic trajectory through the series of exchanges. This will also serve to illuminate the 5th complexity related to power relations in transmodal communications.

'Domestic Work'

After the exchanges above, the Uganda youth finalised and posted a video entitled Domestic Work. The opening shot shows a map of the world, which transitions into a video of their school (from the outside) with the school bell ringing. The next shot is captioned, 'Heading Home After Classes 3:00 pm', and shows the youth walking towards their houses. They are speaking together in Lugbara. The video then transitions to a new screen, captioned 'Time for Domestic Work After Classes'. Two girls leave their home carrying jerry cans (large yellow plastic containers with handles on the side) and walk to a muddy stream. They submerge the jerry cans and fill them with water from the stream. They then twist a rag into a small circle, place it on their heads, and balance the full jerry cans on top of the cloth, carrying them back up the hill to their homes on their heads, hands-free. It is a significant distance. The video transitions to a new scene, music starts, and two girls gather some dirt with a hoe, place it in a tub with (unidentifiable) other matter, and knead it with their hands until smooth. The caption reads, 'Smearing Their Parent's room: Using the Manure of Cow Dung, Soil and Water'. When it is ready, they carry the tub inside and begin to smear the mixture onto the floor of the hut. Another transition, and we see 'The Three Boys Washing Utensils'. For 30 seconds they wash dishes in a washtub (outside on the ground) and scrub pots. Next, they carry a plastic tub and a jerry can of water into a field, and water cows. In the final segment they enter a hut, gather a metal pan, a mortar, a straw basket, and dried cassava, and carry them outside, where they proceed to grind the cassava into cassava flour for cooking then sift it through the basket, all with background music. This is captioned, 'The Four Boys are Pounding Cassava in the Traditional Way'. To end, there is scrolling text, entitled 'Acknowledgement'. It reads, 'Our first appreciation and thanks goes to God for His support and love, indeed without His support our success would be imaginary'. The entire video is 5 minutes long.

In juxtaposition with the Barcelona video, this is a portrayal of very different lives and communities. These youth must indeed do 'domestic chores' after school to survive. The physical space and material objects stand in stark contrast, as does the pace of the video. The seriousness is apparent, as opposed to the playful nature of the other. The modal choices they make are quite different, and perhaps more technically sophisticated (although they, unlike the Barcelona youth, had never used a computer prior to the project start). Excerpts from the chat discussion:

(B) Hello our friends!!! we really liked your video, we are suprised about how different your routine is. We want to ask you some questions:
How do you share the household chores? Do girls always go to take the water?

(U)   Yes girls do fetch water as well the boys. The way we divide household chore, girls in most cases do work which are in line with food preparation, cleaning, while boys may go for farming and grazing.

(B)   Do you do the household chores everyday?

(U)   Yes the household chores are in daily basis.

(B)   Is it hard for you to do the household chores or you are used to?

(U)   Absolutely we are used too, and it has part of our daily work.

(B)   how do you learn to brilng the water on the head?

(U)   Balancing the jerry of water on head is something very easy with the help of head ring, boys too can balance it.

(B)   At what age do you start taking the water on the head?

(U)   As early as five years old.

(B)   What is the mixture for?

(U)   For smearing our room because they are not cemented neither are they tiled

(B)   When the mixture is dry, does it smell?

(U)   After drying it does not smell.

This is a robust exchange of information, with the Barcelona youth demonstrating deep interest in and curiosity about the lives of their global peers. It is clear that the activities and lifestyles of the Uganda youth are unfamiliar to them. Interestingly, remembering that the Barcelona questions were posted as a batch prior to any Ugandan responses, the Barcelona youth added these statements at the end of their original set of questions.

> In our case we have fountains everywhere, so we don't appreciate it. We think you take care more about the water you have and you don't waste it.

> In relation to the household chores, we think that you have a very responsable attitude because we only help at home if our parents ask for. moreover we tend to complain.

This, I believe, demonstrates one of the goals of the project; in considering the lives of global others, youth are provided a lens for self-reflection, and come to understand themselves and their own lives differently. However, in response to the answers from Uganda, the Barcelona youth made and posted a new 49-second video, entitled, 'Trying to Imitate the Ugandan Friends Skills'. The opening shot is text, reading, 'we have tried to imitate your skills'. For 25 seconds the Barcelona project youth, outside on the street, attempt to balance large clear jugs of water on their head (and fail). They are chatting (in Catalan) and laughing, saying, 'look at her!' (in English) and encouraging one another. Then there is a screen that says, 'and we end up doing this…', and they in turn hold the water jug by the handle and throw it up in the air, with a twist of the wrist, to see if they can get it to land on its bottom. There is no explanation other than the two text slides. The Ugandan youth responded with:

> Oh friends, watched your video where you were trying to imitate our skills of carrying water on the. Its so funny and interesting. The two who tried to do so. What was vary hard> because we believe that all ladies can do the balancing not even jerrycan water only but anything that needs to be carried on the head. Why were are you throwing the bottles up?

And the Barcelona youth response:

> We throw the bottles because it's funny and it's a trend.

Transmodal Analysis

So let us once again invoke the transmodalities complexities for analytic purposes.

Complexity #1: Modes intertwined

Youth had two distinct ways to communicate: videos and chats. Meanings were reflexively constructed across both; they seamlessly integrated as youth moved between them. Further, while individual modes are identifiable and carry meaning (written text, speech, music, action, image, pace, and so on), none carried meaning in isolation, and each interacted with others within the multimodal production and flow to create meaning, albeit different meanings for different participants.

Complexity #2: Modes, language and material objects

In this set of exchanges – this semiotic trajectory – modes carried meaning, as did language, but material objects contributed their own semiotic weight. As but one example, the water jug (jerry can) and the water itself mean something entirely different in each site. Their positioning within the youth's lives is different, the youth's relationship to them is different, and their impact on and import for youth's daily lives is different. Water itself, and its carrier, means differently, even in the absence of words and other signs and symbols, and more so (and more obviously so) when entangled with other modes. And this shapes the communications, negotiations and understandings (or misunderstandings) among the youth.

Complexity #3: The arc of communication

As in the first video (from Barcelona), messages that youth intended to convey were often not those taken up. In the Domestic Work video from Uganda, while the activities portrayed were quite familiar (in fact, everyday) for the Ugandan youth, they were quite new to the Barcelona youth, who had no frame of reference for understanding them. However, the project platform enabled questions and answers, leading to negotiations of meaning from which some degree of understandings began to emerge, and were then built on in successive exchanges.

Complexity #4: Context and culture

This set of exchanges created a semiotic trajectory, where the exchanges themselves developed a context within which participants made sense of the various communications and messages. We can see the meanings and understandings shaped in integration with and in counterpoint to the unfolding dialogue. Yet the contexts of the youths' lives – the spaces and places they inhabit, as well as their everyday lifeworlds – shape what they share, how they represent it, and what they can see and understand of others. Further, these interactions are occurring among youth who do not share cultural practices, values, norms and beliefs. Thus it is clear that the Uganda youth are initially unable to interpret the 'sketch' activity ('Revange of the Coins'), having no prior experience with the sorts of texts and play that underlie it, while Barcelona youth do not understand the seriousness of water in the lives of the Ugandans. Their foundational framings do not enable shared understandings of semiotic affordances in communication.

Complexity#5: Relations of power

It is clear that, while both sets of youth live in under-resourced communities, there is not one definition of poverty, and there are vast differences in access to resources and capital. Further, the conditions of life are harsher in Uganda. This plays out in the interactions, at first implicitly, but ultimately (in the final video) rather explicitly. The Barcelona youths' games with water transgress the realities of life for those from Uganda. In the meeting where they discussed the final communication from Barcelona ('it's funny and it's a trend'), they said, 'But for us this is real! This is our life!' Their perceptions of the Barcelona youths' attitudes were hurtful to them, and it was clear to them that there was in fact little understanding in the Barcelona site of their lives and communities.

Conclusion

In this chapter I have called for in-depth explorations of semiotic trajectories in transnational communication and introduced two frameworks that together can provide heuristics for better understandings of encounters between disparate groups of people across distances and time. It is clear that shared foundational ('cultural') assumptions, perspectives and resources facilitate communication. In the absence of these, how can initiatives foster open and caring exchanges and even dispositions that lead to the formation of socially just and equitable relations?

As is clear from the analysis presented, we cannot simply put people – even youth – into contact with one another and anticipate that this will be the outcome. While we can – and should – offer opportunities for direct engagement between people across the globe in scaffolded learning environments, we must attend to issues of misalignment, misunderstanding, and inequitable dialogic interactions. This calls for the following:

- critical reflection and reflexivity (scaffolded) with direct attention to issues of equity and social justice;
- longer-term exchanges in service of developing trusting and sustainable relations;
- transnational collaboration between researchers to gather, share and co-analyze emic perspectives on exchanges and emergent understandings;
- new methods of analysis derived from conceptual framings of transmodalities and critical cosmopolitanism.

This is an ambitious agenda. Yet deep understandings of transmodal interactions, how they shape understandings, and how they foster relationship-building through collaborative transnational engagements, hold the promise of informing the design of critical cosmopolitan educational endeavors – in informal and even formal sites of learning – that may ultimately lead to a more friendly, open and just world.

References

Appadurai, A. (1996) *Modernity at Large: Cultural Dimensions of Globalization.* Minneapolis MN: University of Minnesota Press.

Appiah, K.A. (2005) *The Ethics of Identity.* Princeton, NJ: Princeton University Press.

Appiah, K.A. (2006) *Cosmopolitanism: Ethics in a World of Strangers.* New York, NY: W.W. Norton & Co. Inc.

Barad, K. (2003) Posthumanist performativity: Toward an understanding of how matter comes to matter. *Signs: Journal of Women in Culture and Society* 28 (3), 801–831.

Canagarajah, A.S. (2013) *Translingual Practice: Global Englishes and Cosmopolitan Relations*. New York, NY: Routledge.

Canagarajah, A.S. (2018) Translingual practice as spatial repertoires: Expanding the paradigm beyond structuralist orientations. *Applied Linguistics* 39 (1), 3–54.

Delanty, G. (2006) The cosmopolitan imagination: Critical cosmopolitanism and social theory. *The British Journal of Sociology* 57, 25–47.

Flewitt, R. (2011) Bringing ethnography to a multimodal investigation of early literacy in a digital age. *Qualitative Research* 11 (3), 293–310.

García, O. and Li, W. (2014) *Translanguaging: Language, Bilingualism and Education*. New York, NY: Palgrave Macmillan.

Hall, S. (2006) Interview of Stuart Hall, March 2016. www.youtube.com/watch?v=fBfPtRaGZPM (accessed January 2020)

Hansen, D. (2010) Cosmopolitanism and education: A view from the ground. *Teachers College Record* 112 (1), 1–30.

Hawkins, M.R. (2010) Sociocultural approaches to language teaching and learning. In A. Creese and C. Leung (eds) *English as an Additional Language: Approaches to Teaching Language Minority Students* (pp. 97–107). London: Sage.

Hawkins, M.R. (2014) Ontologies of place, creative meaning making and critical cosmopolitan education. *Curriculum Inquiry* 44 (1), 90–113.

Hawkins, M.R. (2018) Transmodalities and transnational encounters: Fostering critical cosmopolitan relations. *Applied Linguistics* 39 (1), 55–77.

Heller, M. (ed) (1988) *Codeswitching: Anthropological and Sociolinguistic Perspectives*. Berlin: Mouton de Gruyter.

Hull, G.A., Stornaiuolo, A. and Sahni, U. (2010) Cultural citizenship and cosmopolitan practice: Global youth communicate online. *English Education* 42 (4), 331–367.

Jewitt, C. (2011) An introduction to multimodality. In C. Jewitt (ed.) *The Routledge Handbook of Multimodal Analysis* (pp. 14–27). London: Routledge.

Jewitt, C. (2017) *The Routledge Handbook of Multimodal Analysis* (2nd edn). London: Routledge.

Kress, G. (2011) 'Partnerships in research': Multimodality and ethnography. *Qualitative Research* 1 (3), 239–260.

Kress, G. (2017) What is mode? In C. Jewitt (ed.) *The Routledge Handbook of Multimodal Analysis* (pp. 60–75). London: Routledge.

Kromidas, M. (2011) Elementary forms of cosmopolitanism: Blood, birth and bodies in immigrant New York City. *Harvard Educational Review* 81, 581–605.

Latour, B. (2005) *Reassembling the Social: An Introduction to Actor-Network-Theory*. Oxford: Oxford University Press.

Lemke, J.L. (2002) Travels in hypermodality. *Visual Communication* 1 (3), 299–325.

Li, W. (2018) Translanguaging as a practical theory of language. *Applied Linguistics* 39 (1), 9–30.

Nussbaum, M. (1997) *Cultivating Humanity*. Cambridge, MA: Harvard University Press.

Rampton, B. and Charalambous, C. (2012) Crossing. In M. Martin-Jones, A. Blackledge and A. Creese (eds) *The Routledge Handbook of Multilingualism* (pp. 482–498). New York, NY: Routledge.

Rizvi, F. (2008) Epistemic virtues and cosmopolitan learning. *The Australian Educational Researcher* 35 (1), 17–35.

Toohey, K. (2018) *Learning English at School: Identity, Socio-material Relations and Classroom Practice* (2nd edn). Bristol: Multilingual Matters.

Vygotsky, L.S. (1978) *Mind and Society*. Cambridge, MA: Harvard University Press.

2 Translanguaging in ESOL: Competing Positions and Collaborative Relationships

James Simpson

Introduction

This chapter addresses a central issue for researchers, practitioners and policymakers in the field of English for Speakers of Other Languages (ESOL). This is the branch of English language teaching and learning which concerns adult migrants in English-dominant countries. The issue is the extent to which a multilingual or *translanguaging* stance should be supported in ESOL pedagogy. It is prompted by a recognition that competence in English is typically developed by adult migrants as part of a multilingual, multimodal communicative repertoire, used fluidly and flexibly, in spaces that are often heavily multilingual. My discussion is based on the study of activity on an email forum called *ESOL-Research*, an online environment which (following Gee, 2005) I characterise as a *Semiotic Social Space*. This is defined by Gee as an online space (as opposed to a community) which has content (i.e. something for the space to be *about*), and particular ways of organising social interactions in relation to that content.

In this chapter I refer directly to the UK context, though my discussion relates to migrant language education worldwide. The chapter draws upon research carried out during a multi-site ethnographic study of urban multilingualism, *Translation and Translanguaging (TLANG)* project.[1] Findings from TLANG, in common with other work in the sociolinguistics of migration, suggest that the dominant language of the UK is not necessarily the only language to which people need access in their social or work life. The multilingual reality of ESOL learners' lives tends not to be acknowledged in either ESOL policy or practice, however, suggesting an inherent contradiction in the ESOL policy-practice-research nexus between a monolingual approach to ESOL teaching (on one hand) and (on the other) the multilingual experience of ESOL students outside their classrooms. I discuss this contradiction with specific attention to ESOL in the UK, and from my own perspective informed by research and experience. After this introduction, I elaborate on the positions adopted in educational and public discourse towards the use of English in migrant language education vis-à-vis the use of other languages. I follow by describing collaborative relationships between practitioners across the range of ESOL provision, policymakers, and other interested people that led to the development of the Semiotic Social Space of the *ESOL-Research* forum.[2] I then present an examination of an outcome of these collaborative relationships, an e-seminar hosted on the forum whose aim was to bring together, in productive interaction, research findings on multilingualism with practitioner/policy concerns about its relevance to classroom practice. I conclude by reflecting upon how language education policy remains intractably monolingualist and monolingualising, even as the UK's urban – and increasingly its rural – places and spaces become more multilingual.

Monolingualism in Practice and Policy

The TLANG project focused on the ways in which linguistic and other semiotic resources are used in urban contexts where speakers have different proficiencies in a range of languages and varieties. TLANG research has findings in common with other sociolinguistic work on language and literacy practices specifically in migration contexts (e.g. Roberts *et al.*, 1992) and more generally (e.g. Blommaert & Backus, 2011), showing that speakers in such contexts are not confined to using languages separately, unless there are work-related or institutional reasons to do so. In most home and social communication they move fluidly across languages, styles, registers and genres – that is, they *translanguage* – as they attempt to make meaning. Hence there is a growing understanding of the multilingual, indeed translingual, realities of communication in the ESOL students' daily lives, linguistically and culturally diverse as they are. There is consequently greater knowledge of the rich range of language repertoires and other communicative resources – as well as the funds of knowledge (Moll *et al.*, 1992) accrued through life experience – which multilingual adults who are ESOL students bring along with them into any interaction, including, potentially, classroom interaction. Despite this, ESOL teachers only rarely make systematic use of students' full linguistic repertoires in pedagogy, to the extent that some ESOL classrooms are explicitly monolingual 'English Only' spaces where multilingual language practices are viewed as being unconducive to learning and are prohibited.

Practice

Positioning multilingualism as a problem in ESOL echoes a pervasive monolingual ideology evident in language education and in educational spaces in the English-dominant west more generally. The ability to use languages other than English in a repertoire is widely regarded as problematic rather than advantageous to students' learning (Cummins, 2000) and 'something to be overcome' (Safford & Drury, 2013: 70). The monolingual bias in language education is especially evident in approaches associated with Communicative Language Teaching (CLT). Here the goal of L2 learning is often assumed to be native-like mastery, or native-speaker competence. This tendency has its origins in an understanding that the emphasis in language teaching should be less on linguistic form and more about communicative function. This is essentially on the grounds that what learners need to learn is what native speakers actually do with their language in natural contexts of use (Wilkins, 1972; Brumfit & Johnson, 1979).

The notion of the native speaker is heavily contested by sociolinguists (e.g. Rampton, 1990). A sociolinguistically-oriented counter-argument to the perception that native-speaker competence should be the aim of language instruction rests on the idea that people who speak differently from the arbitrary group labelled 'native speakers' are not using linguistic forms that are better or worse: they are just speaking differently. This is conceptually (and ideologically) challenging for many language learners themselves, as well as for their teachers.

Those who regard multilingualism as a problem for language teaching, and who promote a monolingual approach, also echo a much earlier development in language education, the reform movement of the late 19th century. The reform movement was responding to the traditional language teaching approach of grammar-translation, upon which much classroom practice had rested hitherto. Hall and Cook (2012) point to the work of Maximilian Berlitz in particular as exerting a powerful influence on language teaching, all the way to present times. Berlitz' language schools were founded on the principle that the use of students' expert languages must be prohibited in the classroom. Hall and Cook (2012: 275) describe the monolingual principle thus: 'Classes in which students are speakers of a variety of languages, and the employment of native speaker teachers who do not necessarily know the language(s) of their students, created situations in which bilingual teaching seemed to be impossible.' Twenty-first century ESOL pedagogy adopts a monolingual position for much the same reason.

Policy

Like their teachers, the managers of many centres where ESOL is taught also adopt strict policies requiring students to speak no languages other than English in their classrooms, maintaining that 'English Only' is the best approach. The personal, institutional and professional discourses about 'English Only in the ESOL classroom' are redolent of, and indeed partly comprise, ways of speaking about language and migration in public and in policy that help shape the landscape of adult migrant language education. A policy debate in recent years is how English is implicated in the construction of national identity. That is, what is the connection between the English language and the notion of 'Britishness' (Cooke & Peutrell, 2019)? ESOL is part of this debate, one in which migrant language learners themselves frequently find themselves centre-stage.

Despite its very obviously multilingual population, the UK is often represented as a monolingual state, or perhaps as a bilingual one in Wales and Scotland. The association of a British national identity with English is underpinned by a strong 'one nation one language' ideology. Accordingly, in order for British society to be cohesive and stable, its population must share a common language. This ideology is evident not just in the UK: similar discourse is a key feature of nation state-building almost everywhere. And in parallel with elsewhere, in the processes that comprise UK language policy, understanding, using and being tested in the standard language of the new country is not only a proxy for national unity, but also for integration, social cohesion and social mobility (Casey, 2016). Language education is heavily implicated in immigration legislation too: people applying for settlement in the UK are required to pass an English language examination in addition to the *Life in the UK* citizenship test (which itself must be taken in English, Welsh or Scottish Gaelic), and a good deal of effort in the field of ESOL is expended in preparing students for assessments to satisfy the requirements for citizenship and naturalisation.

Multilingual Orientations

It is perhaps not surprising then that language and literacy education for migrants in practice and in policy still distinguishes between the so-called native speaker and non-native speaker, rarely embracing bilingualism or multilingualism, or recognising the purpose of language instruction as the development of a multilingual repertoire. Yet the boundaries between languages in use are fluid and porous, as noted above, and can be associated with an increasingly unclear distinction between native speakers and non-native speakers in practice.

An 'English Only' orientation in language education has not gone unchallenged, on cognitive, affective and pedagogical grounds (Nation, 1978; Turnbull & Dailey-O'Cain, 2009; Cook, 2010). In ESOL and ESOL literacy likewise; and in this field there is also a strong tradition of critical pedagogy. This body of theoretical and practical work is oriented towards equipping students with the communicative tools to enhance their audibility. Advocates of critical ESOL literacy reject monolingualism in L2 literacy teaching, positing that teaching students literacy in the L2 rather than the L1 is unlikely to be effective. Drawing on research into biliteracy they argue that adults learning literacy for the first time learn more effectively if literacy is taught in a language they know (Riviera & Huerta-Macias, 2008). Writing about the US context, Auerbach (1993: 18) suggests:

> The result of monolingual ESL instruction for students with minimal L1 literacy and schooling is often that, whether or not they drop out, they suffer severe consequences in terms of self-esteem; their sense of powerlessness is reinforced either because they are de facto excluded from the classroom or because their life experiences and language resources are excluded.

The teachers who participated in the e-seminar I describe below are employed in the linguistically-diverse contexts where their ESOL students live, work and socialise: for at least some of them, restricting classroom language use to only the 'target language' appears to have even less justification. How they might move towards a multilingual, and – beyond that – towards a translanguaging orientation towards their teaching is the topic of the later parts of the next section.

ESOL-Research and the TLANG E-Seminar

The *ESOL-Research* email forum is an online environment for discussion, a message board, a space for debate and activism, and a cohesive tool for ESOL teachers with an interest in research which impinges on their practice. I instigated *ESOL-Research* in June 2006, as a way of keeping in contact with the teachers and researchers working on a project on which I was employed as a researcher, the *ESOL Effective Practice Project* (EEPP: Baynham, Roberts *et al.*, 2007). The *ESOL-Research* forum is not an online community (although it has characteristics of a community), but rather is a Semiotic Social Space, as defined by Gee (2005). Gee developed the notion of a Semiotic Social Space as a response to the idea of a Community of Practice (Lave & Wenger, 1991; Wenger, 1998), considering it less of a community and more of a loose affiliation. Gee suggests that 'Community of Practice' tends to be used too freely: 'community' has connotations of belongingness and close-knit ties, and is not necessarily a good label for just any group of individuals who happen to have something in common. Moreover, 'community' brings with it notions of membership, which can mean so many different things to different people in different sites that it is not necessarily a helpful idea. Sometimes, Gee argues, we should begin with spaces (virtual or physical) rather than people, when considering forms of social configuration. The defining characteristics of a Semiotic Social Space are that it has some content, something for the space to be *about*, and that there are particular ways of organising thoughts, beliefs, actions, and social interactions in relation to that content. For subscribers to *ESOL-Research*, rather than belonging to a community as members, what links them is the *aboutness* of teaching, learning, policy and research around adult migrant language education, usually in the UK.

The communicative practices associated with *ESOL-Research* respond to this potentially very broad content. At some times the discussion might be about a particular pedagogic issue. In quiet periods the main activity might be the posting of job advertisements or requests for information or advice about a teaching concern. But when there happens to be a particularly contentious policy move that affects adult ESOL, traffic on the forum increases, and posts will tend to offer more in-depth personal and institutional perspectives on policy.

From its foundation to the time of writing a dozen years later, the policy landscape within which the *ESOL-Research* forum is located has changed hugely, and the discussion on the forum has reacted accordingly. In particular, *ESOL-Research* has provided a vital space for the debate and contestation of policy moves that threaten the field. In the early years of the 21st century, ESOL, along with Literacy and Numeracy, was positioned in national policy as an adult basic skill, part of the *Skills for Life* strategy. Since the demise of *Skills for Life* in 2009 the responsibility for coordination of ESOL has been handed from national to local government, there have been successive cuts to funding associated with the promotion of austerity measures, and – as noted earlier – ESOL has become ever more closely tied to immigration policy. *ESOL-Research* has at times become a space for activism: for example the pressure group *Action for ESOL* (http://actionforesol. org/) was founded as a consequence of debate in its space.

The TLANG project and the e-seminar

The prime purpose of the *ESOL-Research* forum is the discussion and debate of research and policy that relates to practice. For the remainder of the chapter I describe a discussion at the intersection of research, policy and practice, an e-seminar that took place on the forum in spring 2017 as part of the activities of the TLANG project. The topic of the e-seminar was 'Translanguaging, superdiversity and ESOL,' and it took as a point of departure two questions for discussion:

(1) What are some of the challenges and opportunities that contemporary diversity might present to teachers and curriculum planners working in the field of English for Speakers of Other Languages (ESOL)?
(2) English might be just one of many languages which ESOL students encounter day-to-day. They may well be developing their competence in a range of varieties of English as part of a multilingual repertoire, and may be translanguaging as a matter of course. How might ESOL teachers and their students address this multilingual reality in their classrooms?

The seminar also touched on the contested issue of 'English Only' in the classroom, and the ideologies informing the competing stances on this matter. As a prompt for discussion, a selection of videos, audio transcripts and field notes from the TLANG project was made available online. The first post of the seminar was a response to these materials by three discussants, ESOL practitioners and researchers Dermot Bryers (English for Action, London), Melanie Cooke (King's College London) and Becky Winstanley (Tower Hamlets College, London). Then the seminar was opened to the 1100 or so *ESOL-Research* subscribers for contributions to the discussion, and it ran for ten days. There were 37 messages in total during the course of the seminar. Despite the relatively small number of actual posts, the seminar was an active event. The website on which the data was presented was viewed 1665 times between January and March 2017 by 398 unique visitors from a globally-spread audience. The top 10 countries where the audience resided were: UK (1040 views); US (116); Japan (72); Italy (57); Canada (48); Australia (43); Germany (39); Finland (39); Belgium (19) and Greece (17).

Discussant response

In their response to the first question, the discussants mentioned emergent findings from their research on the *Diasporic Adult Language Socialisation* study (DALS)[3] and other participatory ESOL work they have been involved in, and how these findings might be usefully incorporated into ESOL teaching and learning (see also Cooke *et al.*, 2018, 2019). The discussants raised important questions about the challenges and also opportunities presented to ESOL teachers and curriculum planners by the *superdiversity* (Creese & Blackledge, 2018; Blommaert & Rampton, 2011; Vertovec, 2010) of students in ESOL classes, not only in terms of ethnicity, country of origin, religion and language but also in their reasons for migrating, their migration trajectories and the class differences which manifest in the different opportunities available for work, housing and other resources. The challenges and opportunities, as they noted, are manifold:

- How can we *meaningfully* include all participants in such a highly diverse single space of the classroom?
- How can we learn about and then draw on the complexity of the backgrounds and lives of our students in such a way that it can inform our teaching?
- How can we find lesson content which will be engaging for all, especially in participatory approaches such as ours which seek to address social and political themes?
- How can we realistically support diverse groups of migrant students to navigate the demands made of them in the workplace, their communities and in their daily lives?
- How can we utilise the sheer range of linguistic resources shared by a group of students in our pedagogical approaches?

They responded to the second discussion question by noting that the UK is witnessing an upsurge of nationalism and anti-migrant feeling, 'much of which is focussed on speakers of languages other than English', and that at such a time it is 'more important than ever to make sure that the ESOL classroom is a multilingual space free from such prejudice, where students feel comfortable expressing themselves in any language or combination of languages without fear of discrimination or recriminations; a multilingual classroom with a robust multilingual pedagogy is undoubtedly something we should be aiming for.' They continue:

> This, however, is easier said than done, partly because of [...] logistical problems [...] and partly because some students and teachers – as well as their employing institutions – hold deep-seated beliefs that English is best taught in a monolingual classroom and that translanguaging will hinder their acquisition of English.

The discussion that ensued demonstrates how the airing of competing positions in a collaborative space enabled, over the period of the seminar, the development of the idea that the multilingual realities of migrants' daily lives present a challenge to the dominant monolingualism of public, educational and policy discourse, and that space for translanguaging can be encouraged in ESOL classrooms.

Multilingual realities of everyday lives

A position adopted by some seminar participants was one whereby ESOL practice should, in some respect, reflect the realities of their students' lives. The everyday communicative experience of ESOL students can be complex, involving interaction with people from a range of linguistic backgrounds, and students' lives are typically led multilingually. The point was made by one participant about the ordinariness of multilingualism, the idea that it is unremarkable: 'I can say from my own experience and that of my siblings that moving between languages in daily life as children was never confusing, and in retrospect it added richness to our expression.' Participants reflected upon the way daily interaction carries on outside classrooms, and upon the repertoire of communicative resources that their students have to hand that might support their learning: 'Multilingual repertoires are certainly a reality for our learners and I think as language teachers, we should promote this as a valuable skill to possess.'

Arguments for 'English Only'

In a seminar about translanguaging in ESOL pedagogy, it might be expected that participants will take a position against 'English Only'. However, opinions were not uniformly opposed to keeping languages separate in the classroom. For one participant the language separation standpoint is crucial for the development of fluency, which entails not relying on 'mental translation'. He described how he personally needed to stop 'the habit of mentally translating spoken Spanish into English' when learning Spanish: 'My fluency jumped a great deal once I overcame this hurdle in Spanish and Portuguese, and it takes a lot of practice.' Participants also invoked ideas of equity and level-playing-field fairness, in defence of an English-only classroom space. That is, if the use of languages other than the target language is allowed in classrooms where there are students with a variety of language backgrounds, there is a risk of excluding some students. Activities involving grouping students by language background to enable classroom communication in expert languages would, in the view of one participant, be divisive: 'You would have strong bonds made between those who share their own language(s) but a division between those who speak different languages.' Moreover, like many ESOL students themselves, some teachers fear that allowing other languages into pedagogy will restrict the already limited opportunities that students might have to practice English.

Resisting 'English Only'

Countering a monolingual view, to require multilingual groups of students who are learning English to only speak English in their classrooms is, in the opinion of one participant, 'neither helpful nor respectful.' For another, denial of multilingualism through the promotion of a monolingual classroom space is a concern because not acknowledging students as speakers of different languages 'creates a sort of belief of the superiority of English over other languages and the superiority of a native teacher over others. It creates a stigma of bilingualism and almost a feeling of disrespect towards other languages and cultures.' One teacher noted that the promotion of a pedagogy that involves students' drawing on their range of linguistic resources has an affective rationale: 'It is good for language learners to feel that they have valuable (other) language resources and experiences they can bring to the classroom and usefully share, even if they are struggling with English.' On a more practical note, as one participant pithily put it, 'I can't see a place for a monolingual institution with monolingual practices in a multilingual world.'

Often stimulated by the initial discussant response, the debate had interesting and useful things to say about how a multilingual ESOL pedagogy might be employed, in an effort to contest the monolingualism of contemporary ESOL, or indeed simply to align with the notion that the teaching of English to adult migrants aims to develop their multilingual repertoires. 'It does not seem to make sense anymore (if it ever did)' said one participant, 'to teach students monolingually how to be (more) multilingual.'

ESOL pedagogy can build on students' multilingual language knowledge, though a teacher's multilingual orientation towards ESOL pedagogy does not have to entail a great deal of other language use. Rather, as one teacher said, it involves 'acknowledging that students know, and have a great deal of fluency in, other languages, and allowing them to draw on, talk about and use this.' A pedagogy that reflects and supports students' experience beyond the classroom walls will lead them to consider their communication in terms of the range of linguistic resources available in the local settings where students will inevitably meet languages other than English. How to connect these local realities to the development of competence in English was explored by one participant, a teacher who distinguished between the process of learning (which might be multilingual) and eventual production, possibly more monolingual: 'Students might be asked to interview local residents, who they might share other languages with, or to photograph the linguistic landscape of their area, and then report back on their findings in English, in some format.' The idea of moving from attention to a range of the linguistic repertoire towards greater attention on English as students gain competence and confidence, was advocated by another teacher who employs the use of students' expert languages to support learning activities. For instance, she supports the selective use of multilingual techniques for pre-writing tasks, aiming for idea-generation, 'which I have found to be beneficial and seem to enable learners to produce better work than they had done otherwise.'

Translanguaging Space, Translanguaging Stance

I now consider translingual pedagogic practices in more detail, and do so by bringing in the notion of stance. Stance is a dynamic evaluation of something (material or conceptual) achieved in ongoing interaction through communicative means (Jaffe, 2009). Teachers' choice of content and material, their methods and techniques and their interactions with their students, together comprise a particular stance, towards teaching, learning and learners. A translanguaging stance might be seen in teachers who incorporate translanguaging into their pedagogy to create a translanguaging space, a social space created by and for translanguaging (Li, 2011; see also García & Li, 2014; Li, 2018). A translanguaging stance can involve more than confronting, in pedagogy, monolingual educational discourses with the multilingual reality. When people are allowed to use a rich range of communicative resources (including their expert languages) rather than being restricted to languages in which they are less proficient, they are more able to negotiate and potentially extend their social identities. Valuing students' full communicative repertoires through the incorporation of translanguaging into pedagogy also involves encouraging students themselves to value those repertoires. A teacher in Scotland who is also a language researcher, put it like this on the forum:

> For me translanguaging allows teachers to recalibrate the value of languages involved. If we want learners to be successful in their learning endeavours and value all their resources, they need to be able to see the value of all the things they can do with language(s). Translanguaging balances the power we often see attached to certain languages by proposing that we value all the communicative repertoires that we can draw on.

She went on to suggest that since discussions of language and power are deeply connected to established language ideologies, translanguaging spaces enable the interrogation of language hierarchies. Another participant also considered the way translanguaging spaces might act as spaces for social justice and for redressing power imbalances, but also supposed that how this might operate in classrooms would be of prime concern for teachers: 'But what do these approaches look like "in practice"? And how successful are they (indeed, how do we define success)? And how can we find and develop empirical evidence across these contexts which can then be shared and adapted/added to/developed/critiqued?'

To end, therefore, I present three descriptions of practice where e-seminar participants – identified by name in this case – appear to be promoting the development of a translanguaging space, and in so doing are enacting their translanguaging stance. First Sheila Macdonald talks about how classes at her organisation, Beyond the Page,[4] are set up.

> We work in a highly-participative model. We also have English-speaking and multilingual volunteers. So the group is highly diverse in fluency and literacy in English. [...] We find that by highlighting and valuing multilingual use, the monolingual English-speaking volunteers discover and begin to appreciate how complex the lives of migrant families are. This re-frames the position of the 'expert speaker' as they observe and experience women using more than one means of communication. The role of the English-speaking group member is to adapt her listening and speaking skills to improve communication – i.e. the focus and responsibility is on the whole group, and the idea of 'language learner' is shared. English speakers begin to develop skills of active listening and waiting whilst others process, and of clearer speech.
>
> This is helpful modelling when other power imbalances are in play – for example, we have a Polish family support worker who finds it very difficult not to immediately translate for Polish women, who tend to rely heavily on her, whilst also showing impatience when others need longer to either get peer support in a home language or work it out in English.

Second, Pauline Moon describes how she supports adult students in their learning on an Art and Design course:

> Development of language and literacy practices is an element of all my work with students [...]. Recently I have used multilingual activities in group sessions on 'Written and spoken reflection',

the aims of which were *to strengthen conceptualisations about reflection* and *to reflect in depth in talk and writing*. The students are a mix of monolingual and multilingual language users.

1. I ask for a volunteer, willing to tell the group some of their reflective thoughts about an aspect of their current art/design project, and to respond to my probing questions. My questions aim to push the student to tell us more, in words, about the depth of their thinking […].

2. I ask the other students to talk in pairs about what they noticed about the conversation between me and the student – and to do this in any language(s) […]'

3. I ask the pairs and the volunteer to feed back to the group. The feedback always includes (but is not limited to) how the volunteer used my probing questions to push their thinking deeper.

4. I ask students to get into threes and replicate the same activity: one person reflects; one is the probing questioner; one is an observer (then they switch roles). Students can do this in any language(s) they wish, and we sort the groups out accordingly […].

Our common ground is that we're all trying to facilitate learning. In relation to students who often find it challenging to express their ideas in English, I think I have noticed that something seems to shift, after a student has used one of their expert languages in the activities. I've heard students talking more clearly, in more depth, and more freely in English. This is why I am interested in multilingual approaches – for me they are about facilitating and easing the learning process, inclusion, identity, claiming and using your voice.

Finally Dermot Bryers, one of the seminar's original discussants, describes a lesson at his own organisation, English for Action,[5] an example of critical translanguaging pedagogy in practice (described in detail in Cooke *et al.*, 2019). Members of the class have chosen *Discrimination on the basis of language use* as their topic:

The issue of people being told to "speak English" in public places emerged because I proposed "languages" as a theme. Students discussed where they spoke their mother tongue (or other expert languages). One student said a friend of hers had been told to "speak English" while speaking Polish to her daughter in the supermarket. The following lesson I showed them a depiction of this story and we discussed it:

My students' stories of this were disturbing. I knew it went on, but one student said she suffers this nearly every day on the bus when she speaks to her relatives on the phone in Indonesian. The class was clear that it was about racism.

This is a mixed level group and my feeling is that we wouldn't have been able to have what felt like an important discussion without using students' other expert languages. The group discussion structured by my questions relating to the image above lasted around 25 minutes. One of the students spoke Spanish and one of my colleagues was able to support her with bits of translation during the discussion […].

Did this help their acquisition of English? I guess it would be almost impossible to say. But having this discussion felt important. After the discussion the students shared their stories in small groups […]. One student told us she had confronted a racist man on the bus and she shouted at him in a mix of English, French and Arabic. Apparently it worked. And she got a round of applause from the class-mates when she told the story. Hopefully some of my students will be better equipped next time a bigot tells them not to speak their first (or 2nd, 3rd etc.) language. This would seem a good outcome from an ESOL class.

Conclusion

Teachers face challenges when they try to incorporate multi-lingualism into their teaching, and to adopt a translanguaging stance towards their practice: few are trained in techniques to do so and many are still told on teacher training programmes that 'English Only' in class is a tried and tested approach. This, as noted, aligns with a pervasive monolingualism in policy and public discourse. Interactions with practitioners such as those in the *ESOL-Research* e-seminar reported here, show conversely that the walls of ESOL classrooms need to be porous: that is, the multilingual concerns of students' lives outside the classroom should closely inform the teaching and learning processes that happen inside. In other words, the purpose of ESOL students' language education is to support their experience of life in a new social context.

This is not least because ESOL classrooms, as sites of multilingual engagement, are also very important sites where identity work is done, as e-seminar participants noted. Through providing translanguaging spaces, ESOL practice might enable students to challenge the limited set of identities imposed upon them by policy and institutionally: they might no longer simply be 'student as potential employee', 'student as test taker', or 'student as aspirant legal citizen'. Indeed a transformative experience should be at the core of migrant language education: what use is it otherwise, if it does not relate to students' concerns and their potential social exclusion in ways that challenge that?

Yet multilingual pedagogies which address the linguistic realities of adult migrant students have been conspicuously thin on the ground. Moreover, as Pöyhönen and colleagues (2018) note, 'even as societies have become more diverse, the language policies which impinge upon adult migrants and their education are – with some exceptions – typically monolingual' (Pöyhönen *et al.*, 2018: 489). Political discourse with a bearing upon language education typically makes no reference to students' multilingualism. Language testing regimes for adult migrants do not acknowledge either the multilingual challenges that students face outside class or the multilingual meaning-making resources that they might have to hand to meet those challenges. While the mutual engagement of ESOL researchers and practitioners that takes place within the Semiotic Social Space of the *ESOL-Research* forum shows that there can be a productive interaction between researchers and practitioners, engagement with policy actors about the benefits of multilingual pedagogy remains more elusive. Embedding multilingualism in practice at grassroots level, coupled with supporting teachers striving to adopt a translanguaging stance, remain the most hopeful ways of promoting multilingualism in ESOL policy.

Notes

(1) Translation and Translanguaging: Investigating Linguistic and Cultural Trans-formations in Superdiverse Wards in Four UK Cities (AH/L007096/1) (TLANG), funded by the Arts and Humanities Research Council. The project was led by Professor Angela Creese, and involved teams in Birmingham, Leeds, Cardiff and London.
(2) The homepage and archives of ESOL-Research are at www.jiscmail.ac.uk and the record of the seminar interaction is available in the archives from January to March 2017. I would like to acknowledge and thank forum members for their responses to the issues raised in the seminar.
(3) The *Diasporic Adult Language Socialisation* study (DALS), funded by the Leverhulme Trust, was led by Professor Ben Rampton and Dr Lavanya Sankaram at King's College, London. The teacher-researchers on the associated ESOL project were Dermot Bryers, Becky Winstanley and Melanie Cooke. I would like to thank them very much for their contribution as discussants.
(4) Bey*ond the Page*, led by Sheila Macdonald, brings women from different backgrounds together to break down barriers of language and cultural differences. www.beyondthepage.org.uk/
(5) *English for Action London* provides ESOL courses for adult migrants in London, and is based on Freirean principles of emancipatory pedagogy. www.efalondon.org/

References

Auerbach, E.R. (1993) Re-examining English only in the ESL classroom. *TESOL Quarterly* 27 (1), 9–32.
Baynham, M., Roberts, C., Cooke, M., Simpson, J., Ananiadou, K., Callaghan, J., McGoldrick, J. and Wallace, C. (2007) *Effective Teaching and Learning: ESOL*. London: NRDC.
Blommaert, J. and Backus, A. (2011) Repertoires revisited: 'Knowing language' in superdiversity. *Working Papers in Urban Language and Literacies* 67. www.kcl.ac.uk/ecs/research/research-centres/ldc/publications/workingpapers/abstracts/wp067-repertoires-revisited-knowing-language-in-superdiversity (accessed January 2020)
Blommaert, J. and Rampton, B. (2011) Language and superdiversity: A position paper. *Working Papers in Urban Language and Literacies* 70. www.kcl.ac.uk/ecs/research/research-centres/ldc/publications/workingpapers/abstracts/wp070-language-and-superdiversity-a-position-paper- (accessed January 2020)
Brumfit, C. and Johnson, K. (1979) *The Communicative Approach to Language Teaching*. Oxford: Oxford University Press.
Casey, L. (2016) *The Casey Review: A review into opportunity and integration*. www.gov.uk/government/publications/the-casey-review-a-review-into-opportunity-and-integration (accessed January 2020)
Cook, G. (2010) *Translation in Language Teaching*. Oxford: Oxford University Press.
Cooke, M. and Peutrell, R. (eds) (2019) *Brokering Britain, Educating Citizens*. Bristol: Multilingual Matters.
Cooke, M., Bryers, D. and Winstanley, B. (2018) 'Our languages': Sociolinguistics in multilingual participatory ESOL classes. *Working Papers in Urban Language and Literacies* 234. www.academia.edu/35839204/WP234_Cooke_Bryers_and_Winstanley_2018._Our_Languages_Sociolinguistics_in_multilingual_participatory_ESOL_classes (accessed January 2020)
Cooke, M., Bryers, D. and Winstanley, B. (2019) 'Our languages': Towards sociolinguistic citizenship in ESOL. In M. Cooke and R. Peutrell (eds.) *Brokering Britain, Educating Citizens*. Bristol: Multilingual Matters.

Creese, A. and Blackledge, A. (eds) (2018) *The Routledge Handbook of Language and Superdiversity*. Abingdon: Routledge.

Cummins, J. (2000) *Language, Power and Pedagogy: Bilingual Children in the Crossfire*. Clevedon: Multilingual Matters.

García, O. and Li, W. (2014) *Translanguaging: Language, Bilingualism and Education*. New York and Basingstoke: Palgrave Macmillan.

Gee, J.P. (2005) Semiotic social spaces and affinity spaces: From the age of mythology to today's schools. In D. Barton and K. Tusting (eds) *Beyond Communities of Practice: Language Power and Social Context* (pp. 214–232). Cambridge: Cambridge University Press.

Hall, G. and Cook, G. (2012) Own-language use in language teaching and learning: State of the art. *Language Teaching* 45 (3), 271–308.

Jaffe, A.M. (ed.) (2009) *Stance: Sociolinguistic Perspectives*. Oxford: Oxford University Press.

Lave J. and Wenger, E. (1991) *Situated Learning: Legitimate Peripheral Participation*. Cambridge: Cambridge University Press.

Li, W. (2011) Moment Analysis and translanguaging space: Discursive construction of identities by multilingual Chinese youth in Britain. *Journal of Pragmatics* 43 (5), 1222–1235.

Li, W. (2018) Translanguaging as a practical theory of language. *Applied Linguistics* 39 (1), 9–30.

Moll, L.C., Amanti, C., Neff, D. and Gonzalez, N. (1992) Funds of knowledge for teaching: Using a qualitative approach to connect homes and classrooms. *Theory into Practice* 31, 132–141.

Nation, I.S.P. (1978) Translation and the teaching of meaning: Some techniques. *ELT Journal* 32 (3), 171–175.

Pöyhönen, S., Tarnanen, M. and Simpson, J. (2018) Adult migrant language education in a diversifying world. In A. Creese and A. Blackledge (eds) *The Routledge Handbook of Language and Superdiversity* (pp. 488–503). Abingdon: Routledge.

Rampton, M.B.H. (1990) Displacing the 'native speaker': Expertise, affiliation, and inheritance. *ELT Journal* 44 (2), 97–101.

Rivera, K.M. and Huerta-Macías, A. (2008) Adult bilingualism and biliteracy in the United States: Theoretical perspectives. In K.M. Rivera and A. Huerta-Macías (eds) *Adult Biliteracy: Sociocultural and Programmatic Responses* (pp. 3–28). New York: Lawrence Erlbaum Associates.

Roberts, C., Davies, E. and Jupp, T. (1992) *Language and Discrimination*. London: Longman.

Safford, K. and Drury, R. (2013) The 'problem' of bilingual children in educational settings: Policy and research in England. *Language and Education* 27 (1), 70–81.

Turnbull, M. and Dailey-O'Cain, J. (eds) (2009) *First Language Use in Second and Foreign Language Learning*. Bristol: Multilingual Matters.

Vertovec, S. (2010) Towards post-multiculturalism? Changing communities, conditions and contexts of diversity. *International Social Science Journal* 61 (199), 83–95.

Wenger, E. (1998) *Communities of Practice*. Cambridge: Cambridge University Press.

Wilkins, D. (1972) *The Linguistic and Situational Content of the Common Core in a Unit/Credit System*. Council of Europe.

3 Belonging, Trust and Relationships: Collaborative Photography with Unaccompanied Minors

Sari Pöyhönen, Lotta Kokkonen, Mirja Tarnanen and Maija Lappalainen

Introduction

How can anyone make sense of the life trajectories and the everyday experiences of unaccompanied minors in Finland, as they find their place under new circumstances? How can researchers build relationships of mutual understanding with both adolescents and their counsellors? How can researchers and counsellors gain and maintain the trust of these young people? How can we avoid intimidating them? How can creative and participatory practice, in this case photography, facilitate collaboration in order to go *beyond language*? With these questions in mind, we embarked on a process of collaborative ethnography (e.g. Lassiter, 2005) in a children's home, known as 'a group home for unaccompanied minors' (henceforth group home). The group home is part of a reception centre for asylum seekers, established in 1991, and located in a rural municipality in a Swedish-dominant region of Finland. We use the official, yet highly problematic term *asylum seeker* in this chapter, because it is how the young people are defined according to Finnish legislation (Finnish Immigration Services, 2019). In case of a positive decision on their asylum claims, they would be granted refugee status.

Our insights derive from long-term partnerships with those in the reception centre and the group home. In this reflective chapter, we describe a photography project co-produced by the reception centre with our research team, for which we conducted a linguistic ethnography, *Jag Bor i Oravais* [I live in Oravais]. Ten unaccompanied minors and their counsellors participated in the project, which took place from October 2015 until November 2016. In this chapter, we unpack our theoretical and methodological choices to describe our conscious aims of collaboration, building relationships, negotiating, gaining and maintaining trust. We also reflect on ethically responsible practices and the challenges of undertaking collaborative research with participants who are experiencing vulnerable lives in liminal spaces.

The research team was made up of four researchers: Sari was responsible for co-organising the photography project with Gustaf, a group home counsellor; Maija took part in fieldwork in the reception centre and the group home; Mirja explored pedagogical and linguistic practices in the local school; and Lotta, joining the project team later on, focused on professional relationships of counsellors and other co-workers in the reception centre.

Guiding Concepts

Belonging is a complex, increasingly used, fluid and flexible concept, as concluded by Lähdesmäki *et al.* (2016). In migration studies, belonging is typically discussed in relation to ethnicity and identity, which according to Anthias (2012: 103), may shift focus away from other social categories, including gender and class, and therefore 'away from the importance of meaning and context as parameters of social life'. In this chapter, we understand belonging as multiple, constructed and contested relationships with people and places. People seeking asylum are living in parallel realities, being *here* and also being *there* (Butler & Spivak, 2007). They are trying to resettle in a new environment – either alone or with their family members – and building new relationships while still awaiting a decision on their asylum claim. At the same time, they maintain relationships with those who remain and they retain memories of the places from which they have been dislocated (Lehtonen & Pöyhönen, 2019). They also experience moments of not-yet-belonging (Anderson *et al.*, 2011) or non-belonging (Holzberg *et al.*, 2018), in situations in which the sense of being part of a community is denied to them. Non-belonging may also be understood as a wilful decision to remain outside a group, and a conscious aim to forget everyone and everything that had to be left behind. These processes of belonging, not-yet-belonging and non-belonging were all present in the minds and lives of the unaccompanied minors we encountered.

Trust is a fundamental concept in studying relationships between asylum seekers and people working with them (including researchers). Moreover, it is an essential element in creating a sense of belonging and relationships, communities and even cohesive societies, as Simmel

(1995) argues. According to Sevenhuijsen (1998) trust needs to be studied in relation to other concepts like vulnerability and dependency. Indeed, in many ways asylum seekers are in vulnerable positions, dependent on others' decisions based on the presumed reliability of their stories (Blommaert, 2001). Trust also involves intimacy, high involvement and reciprocity as Marta Bolognani (2007) points out in her research among and with the Pakistani community in the city of Bradford, UK. In this way, trust and belonging intersect.

In this chapter, we seek to highlight how belonging and inter-personal trust are created and negotiated between the unaccompanied minors, counsellors and us as researchers through photography. We focus on *translanguaging* (García & Li, 2014; Blackledge, Creese *et al.*, 2018; Sherris & Adami, 2018) as a way of understanding how participants draw from their full communicative repertoire – including linguistic and other semiotic resources (e.g. Kusters *et al.*, 2017). We consider how photography allows communication across modes and enables topics that might otherwise stay hidden because of a lack of trust and possibility for reciprocal relationships. For us, trans-languaging practices are not simply multilingual practices, but more like pathways towards epistemic solidarity (Van Der Aa, 2017), conviviality (Rymes, 2014), and socially sensitive active listening (Sabaté i Dalmau, 2018) with young individuals when they are telling their stories about their lived experiences and feelings of forced displacement in spatial and social isolation (Rouvoet *et al.*, 2017).

Context: Working with Unaccompanied Minors in a Group Home

We established contact with the reception centre in Spring 2015, starting our fieldwork in Autumn the same year. At that time over 2,300 unaccompanied minors fled to Finland from war zones and inhuman conditions (Finnish Immigration Service, 2016). Over the course of our research there were over 80 young people in the reception centre, almost all of them were boys aged 14–17 from Afghanistan, Iraq and Somalia.

The adolescents only stayed in the group home until they received the final decision for their asylum claims. If granted asylum, they would then move to another group home and live there until they reached eighteen years of age, which is the age of majority under Finnish legislation. If their case was not yet decided when they turned eighteen they were usually placed in the reception centre with other adults.

Each minor had a personal trustee who did not work in the reception centre and whose role was to act as legal guardian. Counsellors, including other co-workers in the reception centre, were therefore not directly involved in the asylum process. The counsellors felt this to be a relief, as it would have complicated their relationships with the minors.

Individuals working with asylum seekers strive to balance between bureaucracy, organisational loyalty, compassion, and responsibilities (e.g. Eggebø, 2012; Wettergren, 2010). This type of work therefore requires a 'special kind of flexibility' as Berg (2012) puts it. But it also requires the ability to maintain a professional relationship with 'clients'. In our study, we found out that at times it was difficult both professionally and emotionally for the counsellors to support minors in the group home, and, moreover to make the place a home for them. Feelings of 'being at home' would require close and lasting relationships with emotional attachment (Yuval-Davis, 2011). There were also conflicting views among the counsellors about whether they should raise hope or even attempt to develop a sense of belonging for the adolescents in the group home. As one counsellor, responsible for the group home administration, states:

> I think that there is no need to put down roots in this place, it cannot be so... We all have roots in some place ... They have their own family there somewhere ... Okay, if they really put down roots in this place... and then they move from here and when they come over here or call here no one knows these persons... because over here the staff and all the kids are different. ... This might sound strange, but I do not want these young people to get roots in this house. (Interview, August 2015)

This interview took place on the final day of the co-worker's professional career before her retirement, and she defined it as a counsellor's testimony on how to stay in the profession and manage the changing situations in the group home. She told her story to Sari, who herself had lived in a children's home for six years, and thus had personal experience of what it means to grow up under the child protection regime in Finland. It became clear that waiting for a decision, being dependent on others, and, at the same time, trying to fulfil the dreams of their parents or other relatives and continue their lives in the same way as any teenager, was challenging for the group home residents. It was also difficult for the counsellors, who had observed the process multiple times previously, and yet had to carry on in their roles as professional workers. At times it was emotionally demanding to visit the group home because of the realities that both the adolescents and counsellors were experiencing. Our role as researchers was to listen carefully and actively, but also to understand when the time was not right for posing questions or initiating conversations.

At the time of our research it took approximately one year for the unaccompanied minors to receive a final decision on their asylum claims. The atmosphere in the group home changed from one of anxious anticipation to extremes: a cheerful joy when a decision was positive; feelings of devastation when negative. In addition, the legally binding medical age testing procedures caused significant anxiety, as they were considered neither accurate nor fair by the counsellors and by the minors themselves. For example, one Afghani boy, D, was required to move to an adult unit, despite the fact that he knew that he had not yet turned 18. He felt very insecure: too young and too small to live with adult men. The counsellors were concerned about this decision, but they did not have the power to appeal against it.

Ulrika Wernersjö (2015) reported similar findings from her research into feelings of home and belonging among young unaccompanied minors in the Swedish countryside, but with somewhat different interpretations. She concluded that the young people participating in her study did not feel that the relationships with the adults working with them were necessarily close or trusting. Despite most of the minors describing staff members at the group home as 'nice', they did not fully trust the adults working with them. For example, communication between the legal guardians (who usually had several minors to represent) and the minors often involved weekly discussions, but only containing information regarding official matters, such as issues related with schooling. The young people in Wernesjö's study also reported that they seldom kept in contact with their legal guardians after they turned 18 years of age and did not officially need to be guarded by an adult anymore. Wernesjö (2015) states that such distant relationships and, indeed, the lack of lasting relationships may ultimately contribute to feelings of not being at home and of non-belonging.

In Wernesjö's (2015) study the group home was not a 'home' in a traditional sense to the participants. Instead, it was a homely place, in which the participants were encouraged to feel a sense of belonging through relationships and shared activities. As shown in Wernesjö's study and in ours, belonging and trust are negotiated through interactions and they require significant time commitments (Kaukko & Wernesjö, 2017; Turtiainen, 2012; Wernesjö, 2015).

As researchers we had been investigating (forced) migration in Finland for a number of years, from our different disciplinary perspectives (critical sociolinguistics, adult migrant language education, intercultural communication). A number of the reception centre counsellors suggested we focus our research on unaccompanied minors in addition to adults seeking asylum. We considered this suggestion with care, aware that it would be extremely challenging – both methodologically and ethically – to gain the trust of these young people. Bearing these emotional tensions in mind, we sought to ease the pressure on the young people by making methodological decisions, including not to conduct research interviews. Our rationale was that these young people had already had to respond to many kinds of questions posed by police, immigration authorities, social workers, their legal guardians and their teachers. We therefore suggested that we undertake a photography project, which might also enhance the adolescents' wellbeing and, at least, temporarily develop a sense of belonging in their new circumstances, while establishing a common ground for meaning making during the research process.

In the following sections we reflect on the processes of undertaking this project, including the collaborative relationships which developed through photography. Our data are a series of fieldnotes and recordings of conversations between researchers and those involved in the project. In doing this we aim to shed light on aspects of doing interdisciplinary research in contexts of migration with vulnerable young people, offering our insights and foregrounding shared experiences.

Methodology: Collaborative Photography

Working with one counsellor, Gustaf, we started to plan a photography project together. At that time Gustaf was in his early sixties, a Swedish-Finnish bilingual, with Swedish as his stronger language, as he often pointed out. Prior to working in the group home Gustaf had been a professional photographer working in his own studio on portraits and commissioned photographs and taken part in several photography and art exhibitions.

Sari and Maija met first with Gustaf in September 2015. He brought a drawing of a snail – given to him by a very young girl – to the initial meetings. The girl's sister had said that she should not give the drawing to Gustaf because she assumed that he would lose it. Gustaf showed us the drawing with pride, explaining that he had carried it with him at all times ever since. For us, the drawing represented a concrete example of the challenges in gaining trust of a young person who had been let down by adults on multiple occasions before. It also illustrated both the kind of attitude required to build trust and close relationships and the challenges in dealing with the emotional stress that these kind of relationships inevitably present. The snail became a symbol of our collaboration with Gustaf.

During our first meeting, he described his understanding of the photography project and the need to involve both the minors and counsellors. Gustaf had previously taken part in a course on 'empowering photography', a concept developed by Miina Savolainen and registered therapeutically-aligned pedagogic method aiming to empower individuals, groups and communities. We were also aware of this method and knew that it had been used for example in a project with girls living in a children's home (Savolainen, 2008). We wanted to use this approach in our own research project because, as Savolainen explains:

> Empowerment is a process which comes about in social interaction. To function, the method does not require verbal process. The change it produces is often a feeling of intimacy and commitment that stems from the experience of being understood. It is an ability to listen to another human being with deeper concentration and a growing experience of your own ability to show love and respect for the close ones. (n.d., para 6)

We agreed that it was important not only to empower the minors but also the counsellors and to establish a mutual and respectful relationship between them. A common language was not always easy to find in the group home, both literally and metaphorically. Taking photographs together would – we hoped – help the counsellors to understand more about the young people and what they were going through. Our role as researchers was therefore to facilitate, and to listen to both the young people and the counsellors.

Gustaf then introduced the idea of the photography project to his colleagues and the boys, 10 of whom decided to participate. The boys all lived together in a flat, administered by the reception centre and owned by the municipality. This arrangement was a temporary solution as there was no space in the group home itself. It was also a practical solution to include only these ten boys in the project, since Gustaf was also based in the flat.

In the initial project stages we established that the overall aim of the photography project was to offer the adolescents the opportunity to express themselves through visual images, find their own (valuable) things, and make these visible both to themselves as well as to the surrounding community. Our conscious choice of 'empowering photography' shares similarities with the photovoice technique (Wang & Burris, 1997), which has been widely used in collaborative photography projects (e.g. Chase, 2017). Therefore, by giving young participants cameras and asking them to document their lived experiences together with their counsellors, we were seeking to capture a perspective that might otherwise be overlooked (e.g. Chase, 2017; Sanders-Bustle, 2003) by themselves and by others around them.

The group home was provided with two digital cameras. We agreed with Gustaf that the only instruction was for the boys to take photographs together with their counsellors of the places and spaces which were meaningful to them during their time in the group home and local area. There were no 'correct' or 'incorrect' ways of taking photographs, and the boys could to take part in the photography project voluntarily, removing themselves if and whenever they wanted. Gustaf's photographic expertise and his position in the group home led to his role gaining increasing importance. He was able create a space of understanding and respect. But having a personal interest in the topic can lead to challenges in drawing the lines between the personal and professional, as demonstrated here by Gustaf:

> …well, actually he (one of the boys) was home alone, and I did not know how to start taking photographs with him, so we went to clean the sauna. I have learned now [sic] that cleaning the sauna is not part of our (counsellors') duties, but it is the janitor that takes care of it (laughs).

At first Gustaf was slightly concerned about the quality of the photographs. Were they good enough for our research? Were they good enough artistically? We had several conversations about this and encouraged Gustaf to make space for the boys and the counsellors to take photographs in their own style. We came to realise that Gustaf's standards were high, but also that the boys wanted to practice new skills while identifying places and spaces that were meaningful for them. In that sense, our collective aims and aspirations complemented each other.

Over the course of the thirteen months of the project, taking photographs tended to be mainly a part of the boys' leisure activities, although they were also able to take photographs in their local school during the day. Gustaf encouraged the boys and their counsellors to continue to engage. He took boys in smaller and bigger groups, and sometimes on an individual basis, careful of the time available and the boys' moods. Working together to take photographs enabled Gustaf and the boys to discuss their feelings more openly and it seemed to enhance a certain level of emotional disclosure. He would notice how the boys reacted and he would ask about things that they considered good or bad, joyful or upsetting. Photography sessions were also used as a reward, for example if the boys did well in school or behaved according to the house rules. This was in contrast with the ethos of the project, but it was not our place to criticise the practices of the group home.

Data: Creating Belonging and Trust

Over the course of our project, we observed the ways in which the circumstances in the reception centre seemed to change every month with a constant flow of new arrivals and lack of space. During our first visit, one boy slammed his bedroom door in our faces. Sari, conscious of her own personal experiences in a children's home, understood the boy's behaviour. She remembered that when adult visitors came to the home it often felt like being a monkey in a zoo. In a similar vein, we – as researchers – did not belong in the group home and were naturally considered as complete outsiders at the beginning.

We understood the importance of waiting for the right moment, either chatting with the counsellors in the kitchen or following the boys as they played computer games in the living room or watched Arabic and Persian television series. We allowed the boys to act first. The counsellors frequently acted as mediators – asking the boys about their school day or how to translate a particular word. Football was a safe topic, since most of the boys liked it and played football themselves. We never asked anything about their journey to Finland, about their family situation or any other previous experiences. Moreover, the boys never started a conversation about these matters, but they openly talked about their future dreams with a sense of hope and a firm desire to move on in their lives. They were also curious to know about our families and our work as researchers.

Most of the conversations in the group home between the counsellors and the boys took place in Finnish. There was an explicit policy among the staff members to speak in Finnish in the group home to support the boys' language learning. The boys were placed in a Finnish-medium school in the Swedish-dominant municipality. This, according to the reception centre directors, was due to the fact that the Finnish Immigration Services, the national agency coordinating asylum policies, required local authorities to provide schooling of unaccompanied minor asylum seekers in Finnish, even in Swedish-dominant areas. Learning Finnish would enable more opportunities to settle in Finland should a positive decision be received (Pöyhönen et al. [2018] explore the monolingual norms at play in more detail). Even though Finnish was preferred as means of communication, multilingualism was both audible and visual: the boys were using Arabic, Dari, Somalian, Turkish and English with each other, while the counsellors usually talked among themselves in Swedish. The furniture in the apartment was covered with post-it notes naming them in all the languages the boys and the counsellors knew. The boys' language skills were admired by the counsellors and their multilingual interaction was highly respected. The boys were able to rest in their own languages, and they were not forced to use Finnish when replying to counsellors, for example.

Contact with the boys was therefore established slowly. Our monthly fieldwork in the reception centre would involve coffee-table conversations about the project and about life in general, with Gustaf usually taking the lead and showing photographs from his laptop. During one coffee-table conversation, shown below in Excerpt 3.1, we discuss the photographs and how they ·came into being. The conversation takes place three months after we started the project. Coffee is served in the living room by Carita, one of the group home counsellors. Gustaf's laptop is on the table and he is showing photographs and talking to researchers Sari, Mirja and Maija and a graduate student Sirkku about the photography sessions. Two boys, M and H, are playing FIFA16 with Sari's son who is roughly the same age. Gustaf tells us about an Afghani boy, P, to whom he was allocated as a counsellor. P was feeling particularly anxious, as he had been waiting for a decision about his asylum claim for a long time. Gustaf was also nervous, because he knew about the decision, but was not allowed to disclose it because he was not his legal guardian. They had visited the local beach to take some photographs and take their minds off the challenging situation:

Excerpt 3.1: Coffee-table conversation with M about the beach

Finnish: standard font
English: *Italic*

Gustaf: yeah and then we continued the journey and immediately on the beach he started to take photographs of the rocks and the waves

Mirja: clearly he has a different way of looking at things, or I mean, not different, but a special way

Gustaf: [...] and then he took photographs of us, maybe because he was asked. But then he asked me to take photographs of him on the beach and

Sari: were these photographs before or after the decision?

Gustaf: they were before, yes. and I asked him why did he want to go to the beach and he said that he comes here whenever he feels angry, but I also assume that he also comes here when he's sad, or whenever he just feels like it

M: to be alone

Gustaf: to be alone and think and

Sari: *do you have a place where if you feel sad or angry that you would like to go, is there a place, I go to the woods with my dog try to kind of breathe and calm down, do you have such place where you would like to go*

M: *yeah, I'm going beach*

Sari: *yeah*

Sirkku: *beach*

Mirja   *yeah yeah*

Gustaf: *you could also go to the beach I think it's in Oravais it's a very good place because there you can be alone and*

M: *no good*

Gustaf: *not alone?*

M: *yeah*

Gustaf: *you have a friend with you*

M: *no, all of my friends are coming in (the main group home) to see*

Gustaf explains how P started to take photographs of the rocks and waves on the beach, and how he asked Gustaf to take a photo of him while he was sitting on the beach. Mirja comments on the artistic quality of the photographs and the inner feelings they relieved, and Sari asks if the photographs were taken before or after the decision on P's asylum claim. In the middle of the conversation M, a Somalian boy, a fluent speaker of English, joins the conversation, and offers an explanation for why P wanted to go to the beach, suggesting that he wanted to be alone with his thoughts and feelings. This comment invites Sari to ask M, in English, if he too has such places where he can go by himself. She shares her own experiences of the need to be alone in a forest with her dog. M mentions the beach too, but not the local beach, as Gustaf suggests, and he states that being alone is 'no good'. He would rather be with his friends.

The conversation above is a typical example of the conversations we had during the project with Gustaf as mediator and M jumping into the conversation while other boys and counsellors are in the living room listening, knowing what we are talking about, but not necessarily joining in. Yet, there is at least an ephemeral sense of conviviality and belonging. M seems to enjoy being in the company of the researchers and counsellors, but when conversations turn to his own feelings (and turn to Finnish), he appears reluctant to talk, as the following Excerpt (3.2) shows. M, it turned out, was a very good pupil. He liked studying, but did not particularly like the school he went to, because he felt he was a few years ahead of the others. That was likely to be the case, because pupils in the school had very diverse school experiences, if any, before they fled to Finland. M wanted Gustaf to come to the school to take some photographs of him over there:

Excerpt 3.2: Coffee-table conversation with M about school

Finnish: standard font
English: *Italic*

Gustaf: and you had the camera with you because, hmm, but you didn't take that many photographs, I took more, but you were happy that day (pause) or?

Sari: did you have a good day?

M #

Gustaf: you can by the way tell me like my sons say to me at home "you can be quiet, I can speak on behalf of myself" (other adults laughing) so that kind of problem

Sari: did you have a good time? (pause) was it nice when you had Gustaf with you?

M: yeah

Gustaf: yes, but one day you said, when I asked that what is good at school so what did you reply? (pause) you said that everything is bad

M: [it is a] small school here

During the fieldwork and visits that followed, we started to get to know M slightly better. He often came to greet us with a long list of questions about studying Information Technology in higher education. His idol was the physicist Stephen Hawking and he knew several of his quotes by heart. He also showed us video clips from TED talks on language learning strategies and setting life goals: 'a drowning man cannot learn to swim' was one example of a motivational statement he had learned. He was convinced that if he focused on studying he would succeed in the future. We found it inspiring to talk with him. He called himself 'little professor', referring to Mirja and Sari and to his own aspirations to become a scientist:

Excerpt 3.2: The 'little professor'

Finnish: standard font
English: *Italic*

M: I am a little professor

Gustaf: little

M: professor

Gustaf: little professor

M: yeah

Gustaf: yeah

[---]

Gustaf: yes, I heard that you were thinking of this one thing, and I thought to myself that "ahaa, yes". I cannot remember what it was, but I just remember thinking that this is not how every regular little boy would instantly think

M: little Finnish, difficult for me to understand

Sari: yes, but you're not any regular little boy

It was increasingly evident that the photography project had many positive outcomes and created coherence within the group home. Together the boys planned locations and tried various techniques following each others' examples. There were fewer conflicts among the boys in the group home during the project, which the counsellors noticed with pleasure. However, the close relationships between the counsellors, particularly with Gustaf, also reflected a typical group phenomenon where 'outsiders' find it challenging to join the group or feel as part of it. Gustaf recalls a situation with one of the young people, H:

This morning I thought of how thankful I am that this (photograph) turned out like this...this H (watching at H's photo on the screen). I thought that finally I got a kind of photograph where he sits and watches a football game. A Real Madrid game that is. OK, all the boys think that Real Madrid is the best team. But still, I thought that it is a good picture. Yet, nowadays he seems to be somewhat out of the group, alone. And I think that he was a bit jealous that others got so much attention.

By December 2015, just three months into the project, the boys had already taken several hundred photographs, from which they selected two–three each for a shared photo album, adding some lines of thoughts linked with the photographs. Gustaf supported the boys by asking questions about why they chose particular photos. Through this process, the boys were able to reveal their emotions and thoughts. The boys usually chose to write the sentences in English or in Finnish. M added the 'drowning man' sentence to his photographs. P chose two: 'Miksi me olemme tässä maailmassa?' (Why are we in this world?) and 'It hurts when you try to make things RIGHT, and all they can see in you is WRONG.' D, who had to move to the adult unit, wrote about his feelings. H wanted to write that his favourite football team is Real Madrid, which at first might seem superficial. But, through his photographs he could now be seen in the same way as any other adolescent boy, not just an unaccompanied minor, whose presence was quite often gendered, racially-profiled and securitized when with his friends (see also the case of Latino adolescents in Barcelona, Block & Corona, 2014).

The majority of the photographs chosen by the boys and their counsellors were in the form of 'selfies' and group portraits. At first, we were worried about protecting anonymity, but soon we understood that this was exactly how the boys wanted to represent themselves – with their names and faces – and it also aligned with the 'empowering photography' method and existing visual culture (Pink, 2007). Selfies brought us back to the notion of empowerment (see also selfies as a source of empowerment, Gomez Cruz & Thornham, 2015) and for whom this project was designed. However, collectively we made a decision not to reveal participant identity in research papers.

In May 2016, our project team was invited to take part in an exploratory workshop as part of a participatory visual methods seminar in Jyväskylä. Gustaf and Sari presented the project together, with the boys, M leading, giving their own impressions about the project and answering questions posed by academics. In addition, some of the photographs were included in a workshop exhibition. Written feedback from the one-day exhibition was collected, with one of the comments stating: '*A great project. Good boys! It told wonderfully, how words can be marginal if you want to say something meaningful. Thanks and [give us] more!*'

With this highly positive feedback from the exploratory workshop and after the boys received their photo albums, Gustaf and Sari began to consider a larger photography exhibition in the neighbourhood in which the group home was located. This required significant input from Gustaf to curate the exhibition and we researchers wanted to make sure that the boys understood the potential publicity this would bring. In late autumn 2016, two photography exhibitions were organised in two local libraries, allowing easy access for local residents to see the exhibition, and possibly put themselves in the boys' positions.

Gustaf worked very hard on behalf of the boys to create the exhibition, as he had in his previous career. Although the concept of 'empowerment' underpins our project, it is important to consider *whose* vision was represented in the final exhibitions. Although the boys had participated in the project for a number of months, and some of them had already moved from the group home, only three boys attended the exhibition openings. The other attendees were counsellors, local officials and residents. We understood that physical presence might have been overwhelming for some of the boys, and we appreciated how the counsellors protected their wellbeing.

YLE TV news in Swedish reported on the exhibition, and the vice director of the reception centre was interviewed, together with Gustaf and H. We found out from the library workers that the exhibitions had received enthusiastic feedback, with people from the local area recognising many of the places in which the photographs were taken. Moreover, and more importantly, the boys were seen as individuals with names and faces – young people, with fears, joys and dreams. The dual role was heavy for Gustaf and embedded so much emotional work that later on he had to take some time off from work. Could we as researchers have been able to stop this happening? There is certainly an important lesson to be learned for future co-produced projects.

Concluding Discussion

In this chapter, we have critically reflected upon a collaborative and co-produced project and have showed how creative and participatory practice can open pathways to lived experiences of young unaccompanied minors. Working *with* research participants in vulnerable life situations was nothing new to us (e.g. Lehtonen & Pöyhönen, 2019; Pöyhönen *et al.*, 2019). What was new, however, was the collaborative process of meaning making through photography (e.g. Banks, 2001; Pink, 2007) and combining the photographs with tiny fragments of conversations and stories told in social interactions, through which belonging and trust were not self-evident but needed to be frequently negotiated. 'If we try, we can fly', said one boy during the photography project. This way of co-producing research was certainly worth trying.

All of the boys involved in the project have now moved away from the group home, and Gustaf himself has retired. Two of the boys, P and M, have kept in touch, and described the project positively in later conversations with Sari. For others, the project was probably a way of passing time and a distraction from worrying about the outcome of their asylum claims. It might have even been something they did not really enjoy but nevertheless joined in since everyone else – including those they looked up to – was involved in the project too. We cannot therefore honestly say that taking part in the project was a voluntary process for all the participants. There was even an incident during which one of the boys explicitly expressed his disappointment that they had been spending their time wandering in the woods and at the beach instead of learning something useful like Finnish. What we can say though is that the boys and the counsellors were able to collect shared visual memories of the time spent in the group home. Based on our multi-sited fieldwork, we learnt to know that the boys had caring professionals around them both in the group home and in the school.

Linguistic anthropologist Jenny-Louise Van Der Aa (2017) uses the concept of 'epistemic solidarity' to describe the act of creating a space in which one is open to a particular type of story: the story that the participant wants to tell. The long-term partnerships (e.g. Baynham & De Fina, 2016) which we built through our collaborative ethnography as ways of gaining trust enabled us to create such a space. In our experience, the stories were waiting to be told and the spaces needed to be created for the unaccompanied minors to be able to do this in their own time and in their own way. Up until this point, because of their circumstances, it had been a life-saving strategy for the adolescents not to trust anyone (see Turtiainen, 2009). Having positive attitudes towards the people with whom one works is, of course, important. But our project demonstrates that a positive attitude is not enough: building trust and relationships between researchers, young asylum seekers and people working with them is a more complex process, which requires long-term commitment.

References

Anderson, B., Gibney, M.J. and Paoletti, M. (2011) Citizenship, deportation and the boundaries of belonging. *Citizenship Studies* 15 (5), 547–563.

Anthias, F. (2012) Transnational mobilities, migration research and intersectionality. Towards a translocal frame. *Nordic Journal of Migration Research* 2 (2), 102–110.

Banks, M. (2001) *Visual Methods in Social Research*. London: Sage.

Baynham, M. and De Fina, A. (2016) Narrative analysis in migrant and transnational contexts. In M. Martin-Jones and D. Martin (eds) *Researching Multilingualism. Critical and Ethnographic Perspectives* (pp. 31–45). London: Routledge.

Berg, K. (2012) Insha Allah – sosiaalityötä pakolaisasiakkaiden kanssa [Insha Allah – social work among refugees]. In M. Strömberg-Jakka and T. Karttunen Teija (eds) *Sosiaalityön haasteet: tukea ammattilaisten arkeen [Challenges in Social Work]*. Jyväskylä: PS-Kustannus, 14–34.

Blackledge, A. and Creese, A. with Baynham, M., Cooke, M., Goodson, L., Hua, Z., Malkani, B., Phillimore, J., Robinson, M., Rock, F., Simpson, J., Tagg, C., Thompson, J., Trehan, K. and Li, W. (2018) Language and superdiversity: An interdisciplinary perspective. In A. Creese and A. Blackledge (eds) *The Routledge Handbook of Language and Superdiversity* (pp. xxi–xlv). London: Routledge.

Block, D. and Corona, V. (2014) Exploring class-based intersectionality. *Language, Culture and Curriculum* 27 (1), 27–42.

Blommaert, J. (2001) Investigating narrative inequality: African asylum seekers' stories in Belgium. *Discourse and Society* 12 (4), 413–449.

Bolognani, M. (2007) Islam, ethnography and politics: Methodological issues in researching amongst West Yorkshire Pakistanis in 2005. *International Journal of Social Research Methodology* 10 (4), 279–293.

Butler, J. and Spivak, G. (2007) *Who Signs the Nation-state? Language, Politics, Belonging*. London: Seagull Books.

Chase, D. (2017) Empowering Belizean youth through photovoice. *The Qualitative Report* 22 (1), 261–270.

Eggebø, H. (2012) 'With a Heavy Heart': Ethics, emotions and rationality in Norwegian immigration administration. *Sociology* 47 (2), 301–317.

Finnish Immigration Services (2019) https://migri.fi/en/glossary_Glossary, (accessed 24 April 2019).

García, O and Li, W. (2014) *Translanguaging. Language, Bilingualism and Education*. Basingstoke: Palgrave Macmillan.

Gómez Cruz, E. and Thornham, H. (2015) Selfies beyond self-representation: The (theoretical) f(r)ictions of a practice. *Journal of Aesthetics and Culture* 7 (1), 28073.

Holzberg, B., Kolbe, K. and Zaborowski, R. (2018) Figures of of Crisis: The delineation of (un)deserving refugees in the German media. *Sociology* 52 (3), 534–550.

Kaukko, M. and Wernesjö, U. (2017) Belonging and participation in liminality: Unaccompanied children in Finland and Sweden. *Childhood* 24 (1), 7–20.

Kusters, A., Spotti, M., Swanwick, R. and Tapio, E. (2017) Beyond languages, beyond modalities: Transforming the study of semiotic repertoires. *International Journal of Multilingualism* 14 (3), 219–232.

Lähdesmäki, T., Saresma, T., Hiltunen, K., Jäntti, S., Sääskilahti, N., Vallius, A. and Ahvenjärvi, K. (2016) Fluidity and flexibility of "belonging": Uses of the concept in contemporary research. *Acta Sociologica* 59 (3), 233–247.

Lassiter, L.E. (2005) Collaborative ethnography and public anthropology. *Current Anthropology* 46 (1), 83–106.

Lehtonen, J. and Pöyhönen, S. (2019) Documentary theatre as a platform for hope and social justice. In E. Anttila and A. Suominen (eds) *Critical Articulations of Hope from the Margins of Arts Education. International Perspectives and Practices* (pp. 31–44). London: Routledge.

Pink, S. (2007) *Doing Visual Ethnography* (2nd edn). London: Sage.

Pöyhönen, S., Tarnanen, M. and Simpson, J. (2018) Adult migrant language education in a diversifying world. In A. Creese and A. Blackledge (eds) *The Routledge Handbook of Language and Superdiversity* (pp. 488–503). London: Routledge.

Pöyhönen, S., Kokkonen, L. and Tarnanen, M. (2019) Turvapaikanhakijoiden kertomuksia sosiaalisista verkostoista ja kuulumisen tunteista. [Stories of asylum seekers about social networks and sense of belonging] In E. Lyytinen (ed.) *Turvapaikanhaku ja pakolaisuus Suomessa* (pp. 183–204). Turku: Siirtolaisuusinstituutti.

Rouvoet, M., Eijberts, M. and Ghorashi, H. (2017) Identification paradoxes and multiple belongings: The narratives of Italian migrants in the Netherlands. *Social Inclusion* 5 (1), 105–116.

Rymes, B. (2014) *Communicating Beyond Language. Everyday Encounters with Diversity.* London: Routledge.

Sanders-Bustle, L. (2003) *Image, Inquiry, and Transformative Practice: Engaging Learners in Creative Critical Inquiry Through Visual Representation.* New York, NY: Peter Lang Publishing.

Sabaté i Dalmau, M. (2018) Exploring the interplay of narrative and ethnography: A critical sociolinguistic approach to migrant stories of dis/emplacement. *International Journal of the Sociology of Language* 250, 35–58.

Savolainen, M. (2008) *Maailman ihanin tyttö – The Loveliest Girl in the World* (P. Taipale and D. McCracken, trans.). Helsinki: Blink Entertainment.

Savolainen, M. (n.d.) *Empowering Photography.* www.voimauttavavalokuva.net /english/menetelma.htm (accessed January 2020)

Sevenhuijsen, S. (1998) *Citizenship and the Ethics of Care: Feminist Considerations on Justice, Morality and Politics.* London and New York: Routledge.

Sherris, A. and Adami, E. (eds) (2018) *Making Signs, Translanguaging Ethnographies: Exploring Urban, Rural and Educational Spaces.* Bristol: Multilingual Matters.

Simmel, G. (1995) *Hur är Samhället Möjligt? – och andra esseär.* Göteborg: Korpen.

Turtiainen, K. (2009) Kertomuksia uuden kynnyksellä – Luottamuksen rakentuminen kiintiöpakolaisten ja viranomaisten välillä. [Stories on the New Step after arriving in Finland. Trust building between quota refugees and public authorities] *Janus* 17 (4), 329–345.

Turtiainen, K. (2012) *Possibilities of Trust and Recognition between Refugees and Authorities. Resettlement as a Part of Durable Solutions of Forced Migration.* Jyväskylä: University of Jyväskylä.

Valtonen, K. (1998) Resettlement of Middle Eastern refugees in Finland. The elusiveness of integration. *Journal of Refugee Studies* 11 (1), 38–60.

Van der Aa, J. (2017) Senga's Story: The epistemological segmentation of narrative trauma. *Tilburg Papers in Culture Studies* 180. www.academia.edu/33170917/TPCS_180_Sengas_Story_The_epistemological_segmentation_of_narrative_trauma_Sengas_Story_The_epistemological_segmentation_of_narrative_trauma_by_Jef_Van_der_Aa (accessed January 2020)

Wang, C. and Burris, M.A. (1997) Photovoice: Concept, methodology, and use for participatory needs assessment. *Health Education Behavior* 24 (3), 369–387.

Wernesjö, U. (2015) Landing in a rural village. Home and belonging from the perspectives of unaccompanied young refugees. *Identities* 22 (4), 451–467.

Wettergren, Å. (2010) Managing unlawful feelings: The emotional regime of the Swedish migration boards. *International Journal of Work Organsation and Emotion* 3 (4), 400–419.

Yuval-Davis, N. (2011) *The Politics of Belonging: Intersectional Contestations.* London: Sage.

4 The Transformation of Language Practices: Notes from the Wichi Community of Los Lotes (Chaco, Argentina)

Camilo Ballena, Dolors Masats and Virginia Unamuno

Introduction

The presence of Spanish in Argentina dates back to the end of the 15th century, when the first colonists from the Crown of Castile brought it to the homeland of nations who languaged very differently. Nowadays, more than four centuries later, it is still the language of power and prestige, as postcolonial nationalist ideologies and policies of domination imposed the idea of Argentina as a monolingual nation on the social imaginary (Unamuno, 2014). The truth is, however, that at present there are 38 first nations living in Argentina (INDEC, 2012), the Wichi community being one of them. With a population of approximately 50,000 people, Wichi live today in an area that expands from southern Bolivia to the Argentine provinces of Chaco, Formosa and Salta.

In 1987, the Law of Indigenous Communities of the Province of Chaco (Law 3258/1987) recognised the value of indigenous languages as means of instruction in formal education. Until then, Spanish had played a prominent role in the construction of a common and hegemonic national identity, since it was regarded by authorities as what Woolard (2007) terms 'the language of anonymity'. Spanish had the symbolic capacity of crossing over class and ethnic/social frontiers and was seen as an instrument of social cohesion and inclusion. However, as Ballena and Unamuno (2017: 124–125) point out, 'at the same time, indigenous languages were physically and symbolically reduced, relegated to a remote pre-national past, […] treated as folk objects [and] categorized as part of a cultural heritage that [had to] be documented before it disappear[ed]'. This new law also acknowledged the right of indigenous children to be educated in their language.

Bringing bilingual and intercultural education to schools in Chaco was not easy, as non-indigenous teachers did not speak nor understand indigenous languages and there were no teachers among the Wichi. The Centre for Research and Training for Aboriginal Modality (CIFMA) was founded in 1987. Young people elected by their communities and able to speak Qom, Moqoit or Wichi were trained at CIFMA. The first degrees issued at CIFMA enabled their bearers to work as Indigenous Teacher Assistants (*ADA– Auxiliar Docente Aborigen*). Once in schools, apart from acting as translators, they did all kinds of odd jobs for non-indigenous teachers, who often abused them emotionally. The situation started to change in 1995, when the first promotion of Bilingual Intercultural Teachers (*PIB – Profesor Intercultural Bilingüe*) from CIFMA earned a degree that enabled them to act as qualified teachers. According to Ballena *et al.* (2016), for the last decade over one hundred Wichi bilingual educators are currently working at educational institutions in Chaco, either as ADAs or as PIBs.

In 2010, Chaco became, after Corrientes, the second Argentinian province that passed a law to recognise the co-officiality with Spanish of the heritage languages, in this case, Qom, Moqoit and Wichi (Law 6604/10). After that, indigenous educators in Chaco, organised into unions, demanded the creation of positions for CIFMA graduates and greater autonomy in the management of schools in indigenous territories. Their claims were partially attained in 2014 by the Chamber of Representatives, but it was not until March 2017 that the Executive Branch of the Chamber passed Executive Decree 309/2017, which among other prerogatives, established that 50% of teachers at public schools hosting indigenous children should also be indigenous. Yet, bilingual and intercultural education today still needs to overcome great challenges, especially because the contents taught are still based on a common national curriculum that ignores Argentina's plurilingual and pluricultural nature and because most indigenous children at the end of their school years have not managed to master Spanish and cannot access universities or take part in social and professional activities conducted through this language.

In this chapter, we describe the process of designing and implementing the Community Educational Linguistic Project of an indigenous-led primary school in the Wichi community of Los Lotes, (Argentinian Chaco). First, we will provide a brief historical outline on bilingual education in Argentina. Second, we will justify why the proposal had to stem from what in Spanish we refer to as a process of *co-labor* (working together with indigenous communities from symmetrical power positions). Third, we will outline the steps followed, paying special attention to the decisions taken and the lessons learnt.

Understanding Bilingual Education in Chaco

The first forms of bilingual education for indigenous children in Chaco are characterised by what Lambert (1973: 25) described as subtractive bilingualism, a 'form experienced by many minority groups who because of national educational policies and social pressures of various sorts [were] forced to put aside their ethnic language for a national language'. The presence of ADAs in primary classrooms guaranteed the use of native languages and the transmission of native cultural values in schools hosting indigenous pupils. However, the fact that ADAs were few in number and mostly relegated to lower levels (Years 1 and 2), explains why native languages were subtracted as Spanish was taught. The arrival of PIBs made it easier to promote a type of additive bilingualism (Lambert, 1973). As bilingual teachers, PIBs could guarantee education in the native language and ensure that the addition of Spanish as a second language did not block the maintenance and development of children's first language. In this sense, Unamuno and Nussbaum (2017) argue that classrooms managed by a teacher only fluent in Spanish and assisted by an ADA limit the possibilities of bilingual learners to access knowledge; whereas discourse practices in classrooms managed by a PIB provide more open and flexible bilingual modes of interaction and greater opportunities for learning. This is explained by the historical social and linguistic asymmetry existing between non-indigenous teachers and ADAs. ADAs were never allowed to mediate between two languages or two cultures as they were at the service of non-indigenous teachers. Classroom management dynamics progressively changed when PIBs held the authority in the classrooms.

Bilingual and intercultural education (*EIB – Educación Intercultural Bilingüe*) in Chaco also faces the challenge of adapting the common national/provincial curriculum to the needs of indigenous children. The paradox is that indigenous children are expected to reach the common core state standards of a Western based curriculum that sometimes conflicts with heritage knowledge and cultural beliefs of the first nations (e.g. the Western science classification of non-living beings include entities described as living in the Wichi culture).

In this context, indigenous communities see the need to create their own schools and claim the right to manage them. In indigenous-managed schools or in schools in which PIBs outnumber non-indigenous teachers, educational proposals consider how to present conflicting knowledge and how to make use of children's bilingual resources to construct that knowledge. As we will see, empowered

PIBs promote language practices that cannot be categorised as additive, rather, they constitute an example of what García (2009) refers to as dynamic bilingualism in the sense that 'bilingual students and teachers engage in complex discursive practices [...] to communicate and appropriate subject knowledge and to develop academic language practices' (García, 2014: 112). García (2009) uses translanguaging to refer to this pedagogical proposal that accepts as a legitimate practice the dynamic and integrative use of more than one language in the classrooms (see Introduction). The concept is rooted in the belief that bilinguals are 'successful multi-competent speakers' (Cook, 1991: 190–191) able to develop 'communicative expertise' (Hall *et al.*, 2006: 232) by anchoring their discourse practices on their bi/multilingual repertoires (Nussbaum & Unamuno, 2006). Possessing skills in more than one linguistic code permits speakers to switch from one language to another according to the situation (Coste *et al.*, 1997) and 'it is precisely the possibility [bilingual speakers have] of using their multilingual resources what scaffolds the construction of their linguistic competence in the target language(s)' (Masats *et al.*, 2007: 126) and allows them to acquire new field knowledge.

Establishing the Foundations of a Linguistic Educational Project for the Community

Initiatives emerging from the indigenous communities to create self-managed schools are scarce but not rare. For example, in 2013 Wichi leaders in Los Lotes signed an agreement with the Argentine Ministry of Education to ensure that the new school the community had built in the land donated by Rosendo Sosa, their head leader, would be in their charge. Rosendo Sosa primary school opened its doors in 2014 and became the first indigenous-managed primary school in Chaco. A year later, and for the first time in the history of the region, a Wichi teacher had the position of school principal.

The first challenge the community faced related to the school's educational project design. The principal contacted our research group to seek advice on how to create and implement a bilingual educational project. This was particularly innovative because Argentina still lacks a systematic and explicit policy to develop their policies on bilingual intercultural education (Unamuno, 2015). It was also risky as, traditionally, research and educational projects conducted by academics in indigenous communities had led to ethical conflicts due to exploitative research practices: research had benefitted academics (who had gained prestige in the scientific community) but was not considered to have had a positive impact on the community. Consequently, it was essential to clearly define the nature of the relationship to be established between the research group and members of the community. Ballena and Unamuno (2017) proposed that it should empower the indigenous community and that its members should be central to all discussions and collective agreements adopted with regards to the organisation of learning. They use the term, *co-labor* (Leyva & Speed, 2008), to describe their methodological perspective. Collaboration is negatively indexed in the Wichi community as it recalls the asymmetric relationship traditionally established between those academics who, from power positions legitimised by their universities or scientific institutions, were 'helping' or 'cooperating' with a group or community with no power at all. In Spanish, the term *co-labor* rescues the idea of 'working together', from symmetrical but different positions, for a core objective.

In 2015 our team, made up of non-indigenous academic researchers, non-academic Wichi researchers and Wichi teachers, engaged in our first *co-labor* endeavour: the design of a bilingual and intercultural school project which would merge the requirements of the national curriculum and the needs of the Wichi community. Inspiration came from how schools in bilingual communities in Catalonia and the Basque country developed their educational school projects, namely PEC (PEC/*Projecte Educatiu de Centre*). Masats and Noguerol (2016) describe a PEC as a mission statement document created by the faculty of an educational establishment after a democratic, participatory and consensual decision-taking process to (a) determine the school culture and its pedagogical goals and priority actions, (b) develop the syllabi of all school subjects and (c) choose the teaching approach and principles to manage and organise classrooms. The authors also argue that all PECs should be rooted in the territory and have clear linguistic policies or plans of action on how languages and culture would be taught in an integrated manner and used for everyday purposes. This type of procedure aligned with our idea of *co-labor*, so we involved the whole community (researchers, Wichi teachers, community referents, student-teachers, and families) in the process of developing a Linguistic Educational Project for the Community (*Proyecto Lingüístico Educativo Comunitario* or PLEC). In the following sections, we will describe the steps taken to achieve this goal.

Step One: Co-reconstructing the Educational Expectations of Children and Families

Rosendo Sosa primary school opened to host 70 children, 98% of whom were Wichi. Of these, 95% spoke Wichi at home and in the community and 5% both Wichi and Spanish. In this latter case, the children were from families in which a Wichi woman was married to a *Suwele* (a non-Wichi man). Our PLEC had to depart from the immediate context, and our team decided to learn about the expectations of families regarding the new school and the attitudes towards bilingual education of both children and their relatives.

The expectations of families regarding the new school

Between April and September 2014, our team designed interviews which were later administered to the families with children in Los Lotes. These were our findings:

(a) Families expected their children to learn to read and write in Spanish and Wichi. They valued being literate in Wichi, as in the future this would allow youngsters to occupy posts targeted at bilingual speakers.

(b) Families wanted the school to give their children the linguistic and academic competences in Spanish they would need to access secondary education.

(c) Families valued the presence of non-Wichi teachers in the school and our presence as researchers. They encouraged us to create a truly bilingual and intercultural proposal which respected the language and the culture of the community and facilitated children's access to knowledge not always aligned to cultural values and beliefs.

Families' and children's attitudes towards bilingual education

Children from grades 3 to 6 in Los Lotes were asked to express their views on bilingual education through a collective writing game. They were grouped together and given a paper folded in the form of a concertina. Each child had to write a sentence either in favour or against bilingual education and then accordion fold the paper to hide his/her sentence and pass it to another student who had to follow the same procedure. At the end, we had a list of their statements, some of which we reproduce in Table 4.1 below.

During a meeting with families, a Wichi researcher from our team presented the advantages of implementing a bilingual project in Los Lotes to the community. He then gave the floor to the families, who expressed their worries about whether the presence of Wichi teachers in Los Lotes would decrease children's contact with Spanish and hinder learning. As the conversation took place in Wichi, our fellow researcher summarised for us the concerns of the families (Table 4.2).

Table 4.1 Sample of students' views on bilingual education

Student	Original statement	Translation
A	Me gusta el wichi porque es mi lengua. Quiero aprender castellano. Me gustan los maestros wichis porque me escuchan.	I like Wichi because it is my language. I want to learn Spanish. I like Wichi teachers because they listen to me.
B	Voto el castellano, pero quiero aprender a escribir en wichi porque quiero ser maestra.	I vote for Spanish, but I want to learn to write Wichi because I want to become a teacher.
C	El castellano porque es quiero ir al secundario.	Spanish because er I want to go to high school.
D	Suwele lhañhi.	Spanish.
E	Soy bilingüe y me gusta que la escuela sea bilingüe.	I'm bilingual and I like the school to be bilingual.
F	El wichi es mi lengua. Quiero aprender a escribir en wichi y también en castellano porque quiero ser médico.	Wichi is my language. I want to learn to write Wichi and also in Spanish because I want to become a doctor.
G	El castellano me ayudará a defenderme mejor.	Spanish will help me defend myself better.

Table 4.2 Summary of the concerns Wichi families have about the self-managed new school

Original Statement	Translation
A ellos les preocupa una sola cosa. Que los chicos no aprendan suficiente castellano para su futuro. Ellos casi no hablan castellano y no quieren que los chicos no puedan conocer esta lengua. Por eso prefieren que haya docentes no-wichis y que desde el proyecto acompañemos a los docentes wichis para que enseñen también castellano.	They are just worried about one thing, that children do not learn enough Spanish to succeed in the future. They hardly speak Spanish and do not want their children to fail to know the language. Therefore, they prefer to have non-Wichi teachers in Los Lotes and asks us to guide Wichi teachers so that they can also teach Spanish.

Reflections on the educational expectations of children and families

Our analysis of the statements produced by children and the feedback we got from their families allowed us to state that the Wichi community valued both the presence of Wichi and Spanish as languages of tuition. Learning Wichi, particularly mastering its written register, and heritage knowledge were regarded as positive assets by the community. Yet, families also expected that bilingual education would guarantee children's full access to Spanish, the language 'necessary' to access secondary education and to succeed socially.

The statement produced by student G, and reproduced in Table 4.1 above, summarises the Wichi view on Spanish: a tool to 'defend' themselves. Mastering Spanish would give Wichis access to job positions traditionally unavailable for them and would improve their communication with non-Wichi citizens. Non-Wichis are commonly regarded as violent and aggressive individuals, both physically and symbolically, by the Wichi community. Compared to the communication norms in the Wichi society, non-Wichis speak Spanish with a tone that is too high, do not allow their addressees time to think before responding and talk a lot and fast. Not being fluent in Spanish puts Wichis in an intimidating situation when interacting with non-Wichi speakers. Families feel that by mastering Spanish, their children will be able to interact with non-Wichis from a more equal position.

Step Two: Setting, in *Co-labor*, the Foundations of our Community Educational Linguistic Project

Bilingual education in South American countries is often criticised for targeting the elite. Governments develop policies and promote educational programmes to guarantee that non-indigenous children master Spanish and a foreign language but make few efforts to ensure that indigenous children can learn the language of their community and have access to Spanish as a second language in equal conditions. Current educational policies do not establish how heritage knowledge can be part of the official curriculum in indigenous schools, how languages should be taught nor how schools might organise learning. Consequently, school principals take the responsibility of doing so (Fernández *et al.*, 2012) through the design and development of their PLEC.

Masats and Noguerol (2016: 67–68) argue that schools need to face four ideological challenges when designing their linguistic projects:

(a) Define which languages they would teach.
(b) Determine how they would teach those languages in an integrated manner.
(c) Establish procedures to ensure language learning across the curriculum.
(d) Structure learning around meaningful student-centred tasks linked to their participation in the society.

Our team adopted these four premises when, between March and September 2015, we met educational staff from Los Lotes primary school (four Wichi teachers and one non-Wichi teacher) regularly to design, in *co-labor*, their PLEC. As a starting point, we decided that

Wichi and Spanish would be taught in an integrated manner with cultural knowledge from the two communities (the national school community and the Wichi community). We agreed that children would also access curriculum content through their participation in significant and real tasks that could be carried out in either one of the two, or both languages. At that point, we faced two challenges. On the one hand, we wanted to contribute to the empowerment of Wichi teachers, whose competences are often questioned, both in and outside the community. On the other hand, it was necessary to determine how many language spaces to create.

As children, Wichi teachers, like other indigenous people of their generation, had struggled to learn Spanish in non-supportive school environments. Consequently, their competence in Spanish was questioned by non-Wichi people, whose attitudes undermined the trust Wichi families had in them as potential Spanish teachers. It was therefore important to work in *co-labor* with PIBs to resituate and value their bilingual competence. During our meetings we discussed our ideologies regarding which bilingualism is considered 'legitimate' and which is not. For example, in Argentina people who speak Spanish and a foreign language are considered bilinguals, while people who speak Wichi and Spanish are usually not. Our discussions led us to define bilingualism as a socially categorised ability to use languages purposefully and to appreciate the heterogeneous bilingual resources of Wichi speakers, shaped by individual socialisation trajectories. We then had to define learning spaces in which Wichi teachers and children could use their bilingual repertoires to participate in meaningful school projects.

The information obtained from children and their families regarding their educational expectations and needs was taken into consideration in deciding that the PLEC would be developed in three learning spaces: (1) a space for becoming literate in Wichi, (2) a space for mastering communication in Spanish and (3) a bilingual space to access content knowledge. The idea of creating separate spaces for each language within a bilingual learning milieu has been defended in French immersion programmes in Canada (Cummins, 2007) and in bilingual programmes in the US (García, 2009), but it is innovative in the context of indigenous education in Argentina. The notion of space here refers to a regular weekly time period devoted to the attainment of particular learning objectives.

A space for learning Wichi

Our PLEC for Rosendo Sosa school established that there would be a fixed time slot every week devoted to learning Wichi language and culture through activities led by Wichi teachers and, if necessary, by other members of the community. There were two founding principles for this space:

(a) The time slot devoted to the Wichi lhañhi, the Wichi language, would focus on the objective of supporting children to become literate and to learn to create texts of different genres in their language.
(b) During the Wichi lhañhi time slot, children would also acquire knowledge related to the Wichi cosmovision and culture to assist them in appreciating their own culture and the wisdom of their elders. Elder community members would be invited to the school to tell oral stories, therefore transmitting expert cultural heritage knowledge.

A space for learning Spanish

Between July and September 2015, we held workshops with the Wichi teachers in Los Lotes to discuss our methodological approach to teaching Spanish. Because of those meetings, the Rosenda Sosa PLEC included the following:

(a) The time slot devoted to the Suwele lhañhi, the Spanish language, would be the responsibility of Wichi class teachers from Year 1 to 4. In Years 5 and 6, the non-Wichi class teacher would organise learning in this space.
(b) To help children in Years 1, 2, 3 and 4 become aware of the 'norms' of language use in the classroom, teachers would adopt a dress code. During *Suwele lhañhi* time, teachers would wear a smock-frock with a *Wimpala* badge, a badge representing the indigenous flag, stitched on the lapel. During *Wichi lhañhi* time, teachers would wear no badge in their smock-frocks.
(c) In Years 1 and 2, Spanish would be introduced only through the oral medium to ensure pupils had a safe space to start communicating in Spanish. Writing in Spanish would be introduced in Year 3 by engaging children in simple contrastive and metalinguistic tasks to help them to establish links between writing in Wichi and writing in Spanish.

The first Spanish lessons given by Wichi teachers in the lower levels were not successful as children were reluctant to speak Spanish with them and felt shocked and frustrated when the adults addressed them in this language. As language choices are culturally and emotionally bound, we met again to consider a new teaching approach for this space.

(d) The introduction of Spanish in Years 1 and 2 would be done through simulations. To enhance role-acceptance and stimulate children to use Spanish in the classroom, teachers would use a variety of realia to recreate the simulated environment. For example, for a lesson on shopping at the greengrocer's, teachers would bring baskets, fruit, aprons for the shop assistant and ties, hats or other garments for the customers.

A bilingual space for learning content knowledge

The context made it necessary to create two 'monolingual' spaces in the school to guarantee that at the end of their primary education Wichi children could become fluent and literate in Wichi and Spanish. However, we also wanted to sustain our PLEC as 'a dynamic conceptualisation of bilingualism that goes beyond the notion of two autonomous languages' (García, 2009: 53), as we considered that 'the language practices of all bilinguals are complex and interrelated' (García, 2014: 109) and their communicative competence, understood as a combination of knowledge, skills, attitudes and values (Coste *et al.*, 1997) differ from the sort of competence monolingual speakers develop (Lüdi & Py, 2009; Masats *et al.*, 2007). Additionally, a dynamic bilingual programme could go beyond the constraints of our educational milieu.

Wichi could not be the only language of instruction to teach content because of three main challenges: (a) there are almost no content-specific materials in Wichi, (b) Wichi lacks vocabulary related to some curricular field knowledge, and (c) contents in the Argentine curriculum are tailored accordingly to the needs and cultural background of Spanish speaking children. The creation of bilingual spaces, moments in which the use of the two languages would be 'officially legitimised', appeared to be the best strategy to put the national primary school curriculum at the service of Wichi children. We agreed to adopt the following:

(a) Bilingual spaces would be used to promote the learning of content knowledge.
(b) Wichi would mostly be used orally as the vehicular language in those bilingual spaces to facilitate children's access to contents not linked to their own culture. When faced with linguistic challenges, we would rely on the expertise of those community members who knew how to compose new Wichi vocabulary related to disciplinary contents.
(c) The materials used in the bilingual spaces would mostly be written in Spanish and would serve as reference texts. If available, materials in Wichi would be brought to the classrooms.
(d) Content and language learning would be dealt with in an integrated manner, which meant that the choice of language students would

use to conduct the various school projects would depend solely on the requirements of each task. Some tasks would require children to learn to use Spanish to successfully complete them (for example, learning how to make an oral presentation at the Science Fair about medicine plants); other tasks would require children to learn to write in Wichi (for example, learning how to use local plants to produce and record a DIY lotion to kill lice).

Finally, our PLEC established that class teachers would manage the bilingual spaces. As children in the school in Los Lotes were grouped in three classrooms, it was agreed that the Wichi teachers would be the class tutors in lower (Years 1, 2 and 3) and middle (Years 4 and 5) levels and the non-Wichi teacher would be the class tutor of upper (Years 6 and 7) levels.

Step Three: Co-implementing our Community Educational Linguistic Project in Los Lotes

From a socioconstructivist approach, learning is a process of knowledge construction rather than of knowledge transmission (Duffy & Cunningham, 1996), as it is situated and rooted in the social activities carried out by learners (Mondada & Pekarek Doehler, 2004). Accepting that learning takes place through action implies agreeing that knowledge can only be constructed if it is contextualized and can only be transformed and acquired through social interaction (González *et al.*, 2008). Consequently, we all agreed to use project-based learning (PBL) as the methodological approach to implement the decisions taken in the design of our PLEC. As Dooly and Masats (2011: 43) argue, 'PBL is a methodological approach based on contextualised cooperative learning whose implementation fosters the development of learners' cognitive, social and communicative skills through their engagement in the execution of authentic tasks'. This meant that children gained linguistic and content knowledge through the process of being engaged in a series of authentic tasks that resulted in the elaboration of some sort of final output in the form of a text (e.g. an oral presentation, a written formula, etc.) or an object/ product (e.g. a lotion to kill lice) targeted at a real addressee (e.g. the people present at the Science Fair, the community, etc.) with the purpose of attaining a real goal (e.g. preserving heritage knowledge about medicinal plants, in both examples).

Language Choices at the Project Planning Stage

Designing projects structured through realistic goal-oriented tasks that would help children acquire field knowledge and develop cognitive, social and communicative skills while they worked together in the production of a final text (a leaflet, a poster, a book, etc.) or product (a lotion, a commercial transaction, an object, etc.) required making decisions with regards to the language to be used to create those outputs. For example, in their bilingual space, students in Years 4 and 5 learnt to use local plants to produce a lice-killing lotion to be distributed in the fair the community was organising. We agreed that both Spanish and Wichi should be used during the process of learning to identify the plants, to select the leaves to pick and subsequently to cook and distil. Materials would also be available in both languages (children had the science textbook in Spanish and a poster in Wichi about plants in the region).

Children would also need to produce a leaflet with the recipe to create the lotion. Their class teacher suggested that the leaflet should be written in Wichi, but the other teachers thought it ought to be in Spanish, as was usual. As the leaflet was targeted at the members of the community who could read either language we opted to have it written in Wichi. This was firstly because we wanted to 'symbolically' recognise that written Wichi has a role in the community (see Ballena & Unamuno, 2017 for an insight on writing practices in Wichi) and secondly because we could scaffold children's writing processes in Wichi through a meaningful task. Writing in Spanish was also considered, and children used their notebooks to describe the whole process. Again, note-taking had a two-fold objective: children were learning to write in Spanish while acquiring scientific knowledge, but at the same time their work could serve to prove to educational inspectors visiting schools in the area that children were following the national curriculum requirements.

Language Choices at the Project Implementation Stage

As Masats *et al.* (2007: 127) suggest, the language choices of bilingual speakers 'should not be expected to operate within the logic of diglossic practices, instead, they are likely to [...] serve a communicative purpose and contextualise the activities learners co-construct turn by turn'. It is therefore not surprising that during Suwele lhañhi, the space devoted to learning Spanish, the class teacher and the children occasionally engage in a complex plurilingual practice that cannot be assigned either to Wichi nor Spanish, yet it serves to support children's development of communicative skills in both languages, as we can observe in Table 4.3.

The excerpt below was recorded in a class in which pupils in Years 3 and 4 were creating a simulated dialogue between a shop assistant and a customer. The task was used as a communicative practice that

Table 4.3 Plurilingual practices in the Spanish learning space

Original Conversation		Translation	
1. STUDENT:	buenas noches	1. STUDENT:	good evening
2. TEACHER:	FWALA WUCHE? FWALA WUCHE, HOP TOJH INATHAJH?	2. TEACHER:	IS IT DAYTIME? IT IS DAY TIME. IS IT MORNING?
3. STUDENT:	INATJAH	3. STUDENT:	MORNING
4. TEACHER:	KHA, TOJH YUK **buenas noches** HANDE? HAT'E **buenas noches?**	4. TEACHER:	No, YOU DON'T SAY **good evening**, NOCHE, WHAT IS **good evening?**
5. STUDENT:	**buenas noches**	5. STUDENT:	**good evening**
6. STUDENT:	HUNAJH	6. STUDENT:	AFTERNOON
7. TEACHER:	LAPESEY TOJH HUNHAJH, HAT'E?	7. TEACHER:	AFTER THE AFTERNOON, WHAT IS IT?
8. STUDENT:	INATHAJH	8. STUDENT:	MORNING
9. TEACHER:	KHA, TOJH NEL'A INATHAJH LAPESEY HUNAJH LAPESEY HAT'E?	9. TEACHER:	NO, FIRST THE MORNING, THEN THE AFTERNOON AND THEN WHAT?
10. STUDENT:	INATHAJH	10. STUDENT:	MORNING
11. TEACHER:	INA_	11. TEACHER:	MORN_
12. STUDENT:	INATAJH	12. STUDENT:	MORNING
13. TEACHER:	INATHAJH, HUNAJH LAPESEY?	13. TEACHER:	MORNING, AFTERNOON AND THEN WHAT?
14. STUDENT:	INATHAJH	14. STUDENT:	MORNING
15. TEACHER:	KHA CH'AWHIN'UYA OM ISA TOYUK EH... HUNATSI CHE INATHJH HAT'E	15. TEACHER:	NO, LISTEN TO ME CAREFULLY, YOU SAY GOOD EVENING, THEN MORNING WHAT?
16. STUDENT:	**buenos_**	16. STUDENT:	**good_**
17. TEACHER:	**buenos días(cómo se dice la mañana? buenos días)** EY TOJH LUSI?	17. TEACHER:	**good day (how do you say morning? good day)** AND AT NOON?
18. STUDENT:	**buenas**	18. STUDENT:	**good**
19. TEACHER:	TOJH HUNAJH?	19. TEACHER:	IS IT THE AFTERNOON?
20. STUDENT:	**buenas noches**	20. STUDENT:	**good evening**
21. TEACHER:	TOJH HUNAJH **buenas tardes.** EY TOJH HUNATSI?	21. TEACHER:	THE AFTERNOON. **good afternoon.** IS IT THE AFTERNOON?
22. STUDENT:	**buenas noches**	22. STUDENT:	**good evening**
23. TEACHER:	**buenas noches.** KHA IHI LUS IHI TALES. **buenos días, buenas tardes. buenas**	23. TEACHER:	**good evening.** THEY ARE TWO, NOT THREE. **good morning, good afternoon, good...**
24. STUDENT:	**noches**	24. STUDENT:	**evening**
25. TEACHER:	**buenasnoches** [...] NAKHU NE NOM IS	25. TEACHER:	**good evening** [...] NOW LET'S DO IT AGAIN BUT CORRECTLY

would later be exploited at the nearest town market children would visit to buy fruit to prepare a fruit salad. In Table 4.3, the conversation starts after a student proposes an inappropriate greeting to start their simulated dialogue and the teacher engages him (and the rest of students in the group) in a metalinguistic activity to reflect upon how greetings vary throughout the day and how Wichi and Spanish divide days differently.

Learning a language in a bi/multilingual setting allows students to 'learn the forms of the target language while they learn to interpret the social and interactive meaning of the linguistic resources people use' (Masats *et al.*, 2007: 126). Co-constructing knowledge through communicative practices that rely on code-mixing or code-switching mechanisms to negotiate meaning values children's 'multi-competence' (Cook, 1991), proves that plurilingual language practices help them learn better and, consequently, legitimises Wichi teachers as good teachers of Spanish.

Concluding Remarks

This chapter narrated the dialogical process in which a group of non-Wichi and Wichi researchers and primary teachers engaged with the objective of creating together the Community Educational Linguistic Project (PLEC) of the indigenous-led primary school in the Wichi community of Los Lotes, Argentine Chaco. We briefly described how bilingual education has evolved in the area and then focused on the description and analysis of a dynamic bilingual and intercultural programme co-constructed and reshaped in *co-labor* (=jointly from symmetrical positions) by academics and Wichi educators.

Los Lotes Wichi community anticipates that the new indigenous-led school will improve the social, emotional and academic development of indigenous children through the creation of an educational programme with a two-fold objective. On the one hand, it should promote literacy in Wichi and relate heritage knowledge with curriculum content. On the other hand, it aims to help children learn Spanish in a more supportive context than in the past. Consequently, the design of the school PLEC takes the community demands as a starting point to create project-based learning spaces in which each language separately plays a significant role. A Wichi space guarantees the preservation of the heritage culture and a Spanish space guarantees access to the national language culture. Yet, a plurilingual space is also necessary to relate heritage and academic knowledge.

Our study corroborates the value of translanguaging as a learning pedagogy in bilingual educational contexts such as Los Lotes, as it 'legitimises' plurilingual practices as scaffolding mechanisms for constructing field knowledge and for acquiring communicative expertise in the languages in use. Bilingual teachers do not offer learners opportunities to activate two separate codes pupils can mix at will. Instead, Wichi teachers establish a process which enables children to value their multilingual competence as a socially situated learning resource and as a tool to scaffold the learning of both Wichi and Spanish. In our context, the legitimisation of plurilingual practices in both the Spanish and the plurilingual spaces in the school transgresses long-established language prejudices and serves to repair existing inequalities with regards to what Wichi children can do, say or learn in Wichi and in Spanish, and in turn, empowers Wichi teachers as competent educators in either language.

Long-term educational transformations do not stem from good national educational policies or action plans. Positive changes in bilingual educational programmes for indigenous children can only occur through the sustained, reflective and collective actions of bilingual teachers and community members who take the responsibility of designing, implementing and experiencing change together. As a committed team, we became both policy makers and agents of change. Each daily small decision we took collectively generated changes in attitudes, responsibilities, organisation modes and language practices. This is the greatest strength of research and innovation developed in *co-labor*.

Acknowledgements

This chapter would not have been possible without the invaluable help of the Wichi Community in Los Lotes. Our special gratitude to Alejandra Fabián, Amanda Fabián, Luís Fernández, Néstor García and Isabel Anríquez, the teachers who made possible the project described here and who were part of our research team from the start.

References

Ballena, C. and Unamuno, V. (2017) Challenge from the margins: New uses and meanings of written practices in Wichi. *AILA Review* 30, 120–143.

Ballena, C., Romero, L. and Unamuno, V. (2016) Formación Docente y Educación Plurilingüe en el Chaco: Informe de Investigación. Segunda Parte. Unpublished report.

Cook, V. (1991) The poverty-of-the-stimulus argument and multicompetence. *Second Language Research* 7 (2), 103–117.

Coste, D., Moore, D. and Zarate, G. (1997) *Compétence Plurilingue et Pluriculturelle.* Strasbourg: Council of Europe.

Cummins, J. (2007) Rethinking monolingual instructional strategies in multilingual classrooms. *Canadian Journal of Applied Linguistics* 10, 221–240.

Dooly, M. and Masats, D. (2011) Closing the loop between theory and praxis: New models in EFL teaching. *ELT Journal* 65 (1), 42–51.

Duffy, T.D. and Cunningham, D.J. (1996) Constructivism: Implications for the design and delivery of instruction. In D.H. Jonassen (ed.) *Handbook of Research for*

Educational Communications and Technology: A Project of the Association for Educational Communications and Technology (pp. 55–85). New York: Macmillan.

Fernández, C., Gandulfo, C. and Unamuno, V. (2012) Lenguas indígenas y escuela en la provincia del Chaco: El proyecto Egresados. Paper presented at *V Jornadas de Filología y Lingüística, La Plata*, UNLP.

García, O. (2009) *Bilingual Education in the 21st Century: A Global Perspective*. Malden, MA and Oxford: Basil/Blackwell.

García, O. (2014) Countering the dual: Transglossia, dynamic bilingualism and translanguaging in education. In R. Rubdy and L. Alsagoff (eds) *The Global-Local Interface, Language Choice and Hybridity* (pp. 100–118). Bristol: Multilingual Matters.

González, P., Llobet, L., Masats, D., Nussbaum, L. and Unamuno, V. (2008) Tres en uno: Inclusión de alumnado diverso, integración de contenidos y formación de profesorado. In J.L. Barrio (ed.) *El Proceso de Enseñar Lenguas. Investigaciones en Didáctica de la Lengua* (pp. 107–133). Madrid: Ediciones La Muralla.

Hall, J.K., Cheng, A. and Carlson, M. (2006) Reconceptualizing multicompetence as a theory of language knowledge. *Applied Linguistics* 27 (2), 220–240.

Instituto Nacional de Estadísticas y Censos (INDEC) (2012) *Censo Nacional de Población, Hogares y Viviendas 2010. Censo del Bicentenario. Resultados Definitivos, Serie B2, Tomo 1*. Buenos Aires: INDEC.

Lambert, W.E. (1973) Culture and language as factors in learning and education. Paper presented at the *Annual Learning Symposium on "Cultural Factors in Learning"*. 5th Western Washington State College, Bellingham, Washington. https://files.eric.ed.gov/fulltext/ED096820.pdf (accessed January 2020)

Leyva, X. and Speed, S. (2008) Hacia la investigación descolonizada: nuestra experiencia de co-labor. In X. Leyva, A. Burguete and S. Speed (eds) *Gobernar (en) la Diversidad: Experiencias Indígenas desde América Latina. Hacia la Investigación de Co-labor* (pp. 4–59). Mexico D.F.: CIESAS, FLACSO Ecuador y FLACSO Guatemala.

Lüdi, G. and Py, B. (2009) To be or not to be … a plurilingual speaker. *International Journal of Multilingualism* 6 (2), 154–167.

Masats, D. and Noguerol, A. (2016) Proyectos lingüísticos de centro y currículo. In D. Masats and L. Nussbaum (eds) *Enseñanza y Aprendizaje de las Lenguas Extranjeras en Educación Secundaria Obligatoria* (pp. 59–84). Madrid: Síntesis.

Masats, D., Nussbaum, L. and Unamuno, V. (2007) When activity shapes the repertoire of second language learners. In L. Roberts, A. Gürel, S. Tatar and L. Martı (eds) *EUROSLA Yearbook 7* (pp. 121–147). Amsterdam: John Benjamins Publishing Company.

Mondada, L. and Pekarek Doehler, S. (2004) Second language acquisition as situated practice: Task accomplishment in the French second language classroom. *The Modern Language Journal* 88 (4), 501–518.

Nussbaum, L. and Unamuno, V. (eds) (2006) *Usos i Competències Multilingües entre Escolars d'Origen Immigrant*. Bellaterra: Servei de Publicacions de la UAB.

Unamuno, V. (2014) Language dispute and social change in new multilingual institutions in Chaco (Argentina). *International Journal of Multilingualism* 11, 409–429.

Unamuno, V. (2015) Los hacedores de la EIB: Un acercamiento a las políticas lingüístico-educativas desde las aulas bilingües del Chaco. *Archivos Analíticos de Políticas Educativas* 23 (1), 1–35.

Unamuno, V. and Nussbaum, L. (2017) Participation and language learning in bilingual classrooms in Chaco (Argentina). *Infancia y Aprendizaje: Journal for the Study of Education and Development* 40 (1), 120–157.

Woolard, K. (2007) La autoridad lingüística del español y las ideologías de la autenticidad y el anonimato. In J. del Valle (ed) *La Lengua, Patria Común* (pp. 129-142). Madrid / Frankfurt: Iberoamérica and Vervuert.

Part 2: Collaborative Processes

Comment on Part 2: Collaborative Processes

Adrian Blackledge

How we represent ethnography has become as much a part of methodological reflection as the cultural subjects we study. It is more than thirty years since a collection of essays curated by Clifford and Marcus (1986) drew attention to the crisis of representation in ethnographic writing. Geertz (1988) proposed that when we pay closer attention to the process of the representation of ethnography, we learn to read with a more percipient eye. Ethnographies are complex discursive spaces where many voices clamour for expression (Clifford, 1986). Once dialogism is recognised as a mode of representation, the authority of the single voice of the ethnographer is questioned, as is its claim to represent cultures. Ethnographers have questioned the relation of cultural reality to ethnographic expression (Maynard & Cahnmann-Taylor, 2010), arguing that ethnography cannot claim to directly capture lived experience.

MacLure (2003) alerts us to the dangers of realism in ethnographic research. Ethnography commonly presents its narratives as 'true' accounts of 'real' situations, with findings offered as coherent and non-contradictory. But there are no innocent texts, and we need to open up ethnographies to reveal different configurations, interpretations and contradictions. She proposes a move away from representations of social life which create a singular 'reality' (MacLure, 2003). Ethnographers do not merely represent reality, of course, but create it in their outputs. Researchers have sought forms of representation which take these tensions into account. Since the 1990s, ethnography has extended into ethnographic novels, memoirs and biographies. The orthodoxies of classic ethnography are breaking down in the face of experimental representation across genres and media. Ethnographic fiction, for example, is born of the recognition that cultural representations are crafted, and in this sense fictional; they are partial truths structured by relationships of power and history (Jacobson & Larsen, 2014). Instead of trying to present a full explanation of a cultural group or practice, the author of ethnographic fiction aims to evoke cultural experience using literary techniques to craft conventional ethnographic materials – interviews, participant observation, field notes, photographs – into a compelling story. Ethnographic fictions are based on factual research and events, but they are told from a particular point of view, often with the narrator as a character in the story. Geertz (1988: 140) counsels against conflation of the fictional with the false (or, perhaps, in contemporary parlance, with the fake). It can't be right, says Geertz, that we must at all costs call a spade a spade, 'on pain of illusion, *trumpery*, and self-bewitchment' (emphasis added), or that when the literal is lost, so is fact.

Ethnographic poetry is a further creative response to the crisis of representation in ethnography (Prendergast, 2009). Poetry has the potential to bridge disciplines, emerging in (*inter alia*) anthropology, education, urban geography, nursing, psychology and social work. Ethnographic poems as a method of inquiry and representation are often multivoiced, messy texts (Denzin, 1999), in which no single interpretation is privileged. The writing shuttles between description and interpretation, using voice as a means to write for those studied as well as about them. Poems written in response to ethnographic data can mirror slippery identity negotiation processes, question traditional representation of subjectivities, and create a sense of connection with research participants. In one of the chapters in this section, Emilee Moore and Gina Tavares Manuel offer an account of poetry as a rich resource with which to study oral, visual, embodied and spatial modalities. In the ethnographic research described, poetry becomes the material reference point around which shared social meaning is crystallised. The success of the enterprise relies on poetry not as the outpourings of the inspired individual, but as collaborative practice.

Ethnographic poetics are shaped by the anthropological experience, as the poet reflects on field experiences and reframes them through poetry. Furthermore, poetry offers a language that researchers can access when other modes of representation are not fit for purpose (Faulkner, 2009). The poet and social scientist share commonalities in approach: both ground their work in meticulous observation of the empirical world, are often reflexive about their work and experience, and have the capacity to foreground how subjective understanding influences their work. The ethnographic poet is faithful to external socio-historical experience, while reaching beyond the limits of research material to a sense of aesthetics that enhances understandings of the social world. It is in the enhancement of, and elaboration upon, social research outcomes that ethnographic poetry has rich potential. In their chapter in this section of the book, Jane Andrews and colleagues reflect on the collaborative production of poetry in an interdisciplinary research project. Ethnographic poetry can be an analytical or reflexive approach as well as a representational form.

It is a form of inquiry which challenges notions of authenticity, acknowledges complexity and contests the single, unimpeachable account of events (Butler-Kisber & Stewart, 2009). Ethnographic researchers have begun to adopt poetry as a means of representing and responding to research. Ethnographic poems can enable anthropologists to paint social realities in ways that may prove difficult through ethnographic prose (Maynard & Cahnmann-Taylor, 2010). Ethnographic poetry relies on a belief in the ability of poetry to speak to something universal, or to clarify some part of the human condition (Faulkner, 2009). It comes into its own as a means to enrich ethnography when researchers want to explore knowledge claims and write with greater engagement and connection, to mediate different understandings, and to reach more diverse audiences (Faulkner, 2009).

Ethnographic poetry is able to provide insight by resonating with the voices of research participants, by displaying writing about difference and similarities with no easy, determinate answers, and by engaging in the tension between community insiders and outsiders. It demands the use of the senses to convey experiences and practices of other people, and to explain why human beings think, act, and communicate the way they do. The paradox of the research poem is that in order to be useful for research, it must first be a poem (Maréchal & Linstead, 2010). The best examples of ethnographic poems are good poems in and of themselves (Prendergast, 2009). Faulkner (2009: 74) complains of being 'tired of reading and listening to lousy poetry that masquerades as research and vice versa'. Poetry as ethnographic representation is doubly challenged: to be well crafted verse, and to maintain validity in its research results. A poet may write to what is not yet known; an ethnographer writes more to what has already been learned. The ethnographic poet must do both. That is, like the author of historical fiction, the ethnographic poet must be faithful to external experience, while reaching beyond or through it to an equally true, artful reality, a sense of aesthetics that enhances everyday voices rather than diminishing them (Maynard & Cahnmann-Taylor, 2010). Poetry, indeed, is a naturally occurring mode of human speech (Paterson, 2018: 12).

The effects we call 'poetic' occur when speech is made under two conditions: urgency and shortness of time. Language behaves in a material way and, placed under the dual pressures of emotional urgency and temporal constraint, it will reveal its structure and grain. It becomes rhythmic, lyrical and original. Poetry reveals the underlying metrical and intonational regularity of language, and its tendency to pattern its sounds. It reveals the rhythms that dominate the natural phrase and sentence lengths of language, and its narrative and argumentative episodes. Poetry emerges naturally from our speech as the immediate consequence of emotional urgency, and our desire to communicate this urgency by organising and intensifying those natural features of language which best carry it (Paterson, 2018: 13). Jakobson (1960) argued that there were six functions of language: referential, phatic, metalingual, conative, emotive and poetic. The functions he identified are not separate, but co-exist in shifting hierarchies of dominance: 'The diversity lies not in a monopoly of some one of these several functions but in a different hierarchical order of functions' (Jakobson, 1960: 353). Whilst the predominant orientation of many messages may be to the referential function, 'the scrutiny of language requires a thorough scrutiny of its poetic function' (1960: 356). When we pay attention to repetition and 'sameness' in a text (including an oral text) we are able to make comprehensible its poetic structure. Any noticeable reiteration of the same grammatical concept becomes an effective poetic device (Jakobson, 1985: 42). Repetitions such as parallelisms, whether based on sound, on grammatical categories or on lexical categories, are a natural result of the raising of equivalence to the constitutive device of the sequence. Such structures are identifiable in everyday narrative speech. Through attention to structures of repetition we are able to hear more clearly the voices of speakers in everyday contexts, and to make more visible the social relations between them.

Hymes (1981) applied Jakobson's analysis of equivalence to Northwest US Indian language folk tales collected and transcribed by researchers. The folk tales were somewhere between Jakobson's characterizations of 'poetry' and 'ordinary speech'. In his analysis, Hymes demonstrated that oral narratives are 'organized in terms of lines, verses, stanzas, scenes, and what one may call acts' (1981: 309). Analysis of the poetic structure of narrative 'will add to understanding of language itself and contribute to the many fields of inquiry for which the use of language in telling stories is a part' (Hymes, 2003: viii). Discovering lines and relations in narrative can lead to understanding and interpretation otherwise not possible. We can recognise artistry and subtleties of meaning otherwise invisible. For a true account of the human capacity for verbal art, this is crucial (Hymes, 2003: 96). For example, analysis which highlights the poetic structure of (a segment of) discourse in a city centre meat and fish market brings to light the poetics of everyday speech (Blackledge et al., 2016). In their chapter, Moore and Tavares Manuel present the latter's powerful poem, 'Bleach'. It is not essential to be steeped in Hymesian ethnopoetics to understand the resonance of repetition here. Repetition of the word 'skin' in eight successive lines tells of experiences of racism, of pride, of beauty. The poetry is not only in the message of the poet, but in the aesthetic. By the end of the chapter Tavares Manuel's poetry has become reflexive, representing the process of ethnographic poetry, again repetition the artistic device which drives the poem's momentum: 'Crossing cultures / Crossing roads / Crossing styles / Crossing minds'. Not only is social practice engaged in 'crossings', but so is the representation of that practice as ethnographic poetry.

In addition to structures of repetition and equivalence, the poetry of everyday life is also evident in its rhythm. Language is hopelessly rhythmic because we are hopelessly rhythmic (Paterson, 2018: 341). We start with the rhythm of the self – the heart, and respiration – and attend to the rhythm of the other. Wherever there is interaction between a place, a time, and an expenditure of energy, there is rhythm (Lefebvre, 1992/2015). Blackledge and Creese (2019) offer a poetic interpretation of ethnographic observation of a city centre market. Rhythms are present in the social life of the market – the rhythm of commerce, the rhythm of labour, the musical rhythm of butchers shouting their wares. Rhythms are the music of the market, and the music of the city (Lefebvre, 1992/2015: 45). Rhythm shapes, and is shaped by, the everyday poetry of the market. In the example from Moore and Tavares Manuel it is as much the repetition as the metre that provides rhythm in the poem. Indeed it is this alliance of rhythm and rhyme that is the potential of ethnographic poetry. For Heaney (2002) what all humankind has known and experienced is potentially available through the poem. Heaney's argument (re)connects poetry with ethnography. Ethnographic poetry has the potential to carry utterances away to meanings beyond themselves, to what humankind has known and experienced, to the human condition (Heaney, 1988). The poem is not an end-point or conclusion to the field work or analysis. It overlaps with other texts which represent observed practice. Through an aesthetic sense the poem reaches for an enhanced view, an expanded understanding of practice, and complements other ways of constructing meanings of social life.

Poetry is far from being the only creative medium through which ethnographers are expanding and elaborating upon interpretations of social life. In this section of the book we are provided with accounts of creative meaning making which promise much in the furtherance of collaboration between artists and students of everyday practice. Jessica Bradley and Louise Atkinson describe a programme of collaborative arts-informed pedagogical activities, through which they develop new understandings of multimodal translanguaging practices. The authors consider the transformational affordances of arts-informed inquiry, and discover how innovative ways of working collaboratively offer new opportunities for translanguaging spaces to be created. Bradley and Atkinson consider the pedagogical opportunities provided by arts-based research. Their approach is both arts-based, and arts-informed. That is, the arts are integral to ethnographic study from start to finish. Joëlle Aden and Sandrine Eschenauer propose an enactive-performative pedagogy for language education. As evidence for the efficacy of such an approach they refer to theatre workshops co-led by artists and language teachers. The collaborative approach works with the local school ecology of languages, taking into account students' voices, and acknowledging their capacity to navigate between languages and cultures. The authors consider the transformative power of performance, as students are encouraged to play with languages and embodied communication, and eventually to put on a public performance. Teachers integrate theatre sessions into their pedagogical objectives, and teachers' objectives become the inspiration for theatre sessions. This reciprocity pays dividends in the development of performative pedagogy. Aden and Eschenauer conclude that cooperation between artists, teachers, educators and researchers has rich potential in the rehabilitation of the sensory and emotional dimensions of language teaching and learning. Andrews *et al.* consider arts-based collaborative research processes in the context of a large, interdisciplinary research project. They describe the translation of research data and findings into live performance, through collaboration between academic researchers and a playwright. In their project creative approaches were built into the research from the outset, affording innovative ways of engaging with ideas.

Bourdieu (1999) refers to the novels of Faulkner, Joyce, and Woolf in proposing that social research must relinquish what he terms the single, central, dominant, 'quasi-divine', point of view that is too easily adopted by observers, and also by readers. Instead, he says, in ethnographic research we must work with 'the multiple perspectives that correspond to the multiplicity of coexisting, and sometimes directly competing, points of view' (1999: 3). Arts-based approaches to social research which involve genuine collaboration between academic researchers and creative artists offer immense potential, not only for the dissemination of research findings to wider publics, but also for the generation of knowledge from diverse and multiple perspectives. Poetry, prose fiction, dance, theatre, music, song, collage, sculpture, graffiti, comedy, and many other creative media, are relatively unexplored as means of public engagement, and as ways of elaborating and extending the meanings of social research. The chapters in this section offer excellent starting-points from which to tap this potential.

References

Blackledge, A. and Creese, A. (2019) *Voices of a City Market.* Bristol: Multilingual Matters

Blackledge, A., Creese, A. and Hu, R. (2016) The structure of everyday narrative in a city market. *Journal of Sociolinguistics* 20 (5), 654–676.

Bourdieu, P. (1999) *The Weight of the World. Social Suffering in Contemporary Society*. Stanford: Stanford University Press

Butler-Kisber, L. and Stewart, M. (2009) The use of poetic clusters in poetic inquiry. In M. Prendergast, C. Leggo and P. Sameshima (eds) *Poetic Inquiry: Vibrant Voices in the Social Sciences* (pp. 3–12). Rotterdam/Boston: Sense Publishers.

Clifford, J. (1986) Introduction: Partial truths. In J. Clifford and G. Marcus (eds) *Writing Culture. The Poetics and Politics of Ethnography*. Berkeley: University of California Press.

Clifford, J. and Marcus, G. (eds) (1986) *Writing Culture. The Poetics and Politics of Ethnography*. Berkeley: University of California Press.

Denzin, N.K. (1999) Two-stepping in the '90s. *Qualitative Inquiry* 5 (4), 568–572.

Faulkner, S. (2009) *Poetry as Method. Reporting Research Through Verse*. London/ New York: Routledge.

Geertz, C. (1988) *Works and Lives. The Anthropologist as Author*. Cambridge: Polity Press.

Heaney, S. (2002) *Finders Keepers. Selected Prose 1971–2001*. London: Faber.

Heaney, S. (1988) *The Government of the Tongue*. London: Faber.

Hymes, D. (1981) 'In Vain I tried to Tell You'. *Essays in Native American Ethnopoetics*. London: University of Nebraska Press.

Hymes, D. (2003) *Now I Know Only So Far. Essays in Ethnopoetics*. London: University of Nebraska Press.

Jacobson, M. and Larsen, S.C. (2014) Ethnographic fiction for writing and research in cultural geography. *Journal of Cultural Geography* 31 (2), 179–193.

Jakobson, R. (1960) Linguistics and poetics. In T. Sebeok (ed.) *Style in Language*. Cambridge, MA: MIT Press.

Jakobson, R. (1985) *Verbal Art, Verbal Sign, Verbal Time* ed. K. Pomorska and S. Rudy. Minneapolis: University of Minnesota Press.

Lefebvre, H. (1992/2015) *Rhythmanalysis. Space, Time and Everyday Life*. London: Bloomsbury.

MacLure, M. (2003*) Discourse In Educational And Social Research*. Buckingham: Open University Press.

Maréchal, G. and Linstead, S. (2010) Metropoems: Poetic method and ethnographic experience. *Qualitative Inquiry* 16 (1), 66–77.

Maynard, K. (2009) Rhyme and reasons: The epistemology of ethnographic poetry. *Etnofoor* 21 (2), 115–129.

Maynard, K. and Cahnmann-Taylor, M. (2010) Anthropology at the edge of words: Where poetry and ethnography meet. *Anthropology and Humanism* 35 (1), 2–19.

Paterson, D. (2018) *The Poem*. London: Faber.

Prendergast, M. (2009) Introduction: The phenomena of poetry in research: "Poem is what?" Poetic inquiry in qualitative social science research. In M. Prendergast, C. Leggo and P. Sameshima (eds) *Poetic Inquiry. Vibrant Voices in the Social Sciences* (pp. xix-xlii). Rotterdam/Boston: Sense Publishers.

5 Translanguaging: An Enactive-Performative Approach to Language Education

Joëlle Aden and Sandrine Eschenauer

Introduction

In this chapter, we describe the concept of translanguaging (*trans langager*) at the intersection of Francisco Varela's enaction paradigm (1999) and the performative turn in cultural studies (Fischer-Lichte, 2004; Sting, 2012). We will shed light on two processes that unfold in relation to each other: empathising and living aesthetic experiences that 'anchor abstract knowledge in a sensitive and embodied knowledge of the world' (Aden, 2008: 11). Linking the biological roots of language (Maturana & Varela, 1987) and the aesthetic roots of poetic languages (Lecoq, 1997), we pave the way for an enactive-performative pedagogy for languages (Aden, 2017a; Aden & Eschenauer, 2014). From here we map out language education within a framework of *translangageance* that we define as the process of emergence of a common language that makes sense, through all forms of language (Eschenauer, 2017).

First, we present *translanguaging* (*translangager*, Aden, 2012) within Varela's paradigm of enaction, then we introduce the performative approach that allows us to implement an enactive pedagogy. We go on to illustrate our translanguaging model with a study led by the second author carried out in a plurilingual and pluricultural lower secondary school in the suburbs of Paris. We followed a cohort of 20 students aged 11 to 14 over a period of four years (2011–2015). We assessed the impact of theatre workshops co-led by artists and teachers and used as part of the students' language education. This collaborative approach builds on the local ecology of languages (home, foreign and school languages) and takes into account the students' voices and their capacity to navigate between all their languages and cultures. We show how the mechanisms of empathy (kinesthetic, emotional and cognitive), which engage all forms of language (corporal, cultural, linguistic) interwoven with aesthetic experience, form the very essence of *translangageance*. We will see that this pedagogical approach leads to a new configuration of human relations and social bonds.

Theoretical Framework

Anchoring translanguaging in the neurophenomenological paradigm of enaction

We walk in the footsteps of other researchers who use the notion of translanguaging (Canagarajah, 2011; Creese & Blackledge, 2010; García, 2012) since we create learning environments that do not partition languages; and our work draws heavily on Varela's enaction paradigm (1993, 2017). García, who also uses the notion of languaging coined by Maturana and Varela, brings the 'autopoietic organisation of languaging across national, socio-political, and social interactions' to the fore (García & Li, 2014: 203). Varela himself does not present language as autopoietic. For him, autopoiesis describes a system at its cellular level. He believes it is a mistake to confuse levels of organisation, as each level possesses its own organisational characteristics (Varela, 2002: 175). He argues that language and thought are both the result and the means of a 'structural coupling' between brain, body and environment. Within his enaction paradigm, language and thought (consciousness) emerge in action from 'the embodied mind,' or to put it differently, what he calls the mind 'is nothing but the moving body' (2002: 174). However, we feel the word *enaction* does not sufficiently reflect the complexity it covers. Varela explains this word choice by saying: 'We propose as a name the term *enactive* to emphasise the growing conviction that cognition is not a representation of a pre-given world by a pre-given mind but is rather the enactment of a world and a mind on the basis of a history of a variety of actions that a being in the world performs' (Varela *et al.*, 1993: 9). Within embodiment theories, Varela asserts that our knowledge does not reflect a pre-existing world, but that through action and language we co-determine our environment as much as our environment co-determines us. We are therefore part of a holistic, embedded and dynamic conception of knowledge, which 'depends on a world that is inseparable from our bodies, languages and cultural histories – in other words, from our corporeality' (Varela, 2002: 210).

A growing body of research within the field of cognitive neuroscience suggests the brain's plasticity enables it to adapt to all bodily constraints of its environment. This neuroplasticity also concerns language, as shown by cognitive psychologist Ellen Bialystok in her studies on bilingualism. Her results indicate that the use of more than one language changes the structure of the brain, but not necessarily in the way language teachers might expect. Bilinguals have lower formal language proficiency (smaller vocabularies and weaker access to lexical items) in each of their languages compared to monolinguals. Bilinguals, however, develop a neuroplasticity that increases executive control in nonverbal tasks requiring problem-solving (Bialystok *et al.*, 2012). These results are consistent with our field observations of artistic activities conducted in multiple languages. We find this helps develop empathy and psycho-social cognitive skills, which are the bedrock of linguistic components of learning (Aden & Eschenauer, 2014). Bialystok *et al.* (2012) present a distributed model in which both languages are always active to some degree. For them, bilingualism does not involve a switch from one language to another, but rather a joint activation regulated by executive functions such as attention, inhibition, memory or flexibility, that influence the choice of language depending on the context and the subjects' individual characteristics. Therefore, the change from one language to another within the context of learning new languages is rooted in a shared experience that is lived, embodied and emergent.

(Trans)langager and (trans)langageance: Our perspective

It is necessary to bear in mind that enaction is not a theory of language, it is a paradigm of knowledge within which language holds a central place since it is what connects us to others, to the world and to one's self. (Aden, 2017a). While the English language uses a single word, *language*, in French there is a significant distinction between the terms *langage* and *langue*. In our understanding, the concept of language (*langage*) according to Maturana and Varela describes abilities that make it possible to act together. Language is integrated because it is the result of sensorimotor behaviours produced in social coupling, and it is from this social coupling that shared meanings are born. It encompasses all modes of interaction (bodily, emotional, verbal and cultural) and cannot be reduced simply to verbal languages (*langues*). For Varela and Maturana, 'since we exist in language (*langage*), the domains of discourse that we generate become part of our domain of existence and constitute part of the environment in which we conserve identity and adaptation' (Maturana & Varela, 1987: 234). Thus, we consider that learning new languages (*langues*) entails not only a reorganisation of surface linguistic knowledge, but a dynamic reconfiguration of the entire language system (*langage*), and therefore requires a holistic pedagogy.

Our languages (*langues*) reflect differences in the way cultural groups enact reality. In our approach, the prefix *trans-* highlights the co-creation of shared meaning across cultural differences through actions. It refers to that which goes through, goes beyond and connects the subjects to their environment, to others and to themselves. It reflects the complex dynamic which is found in the *spaces between* the physiological and the mental, between oneself and others, between the known and the foreign – with all foreignness being a part of oneself (Sting, 2012). Within a homogeneous cultural group, languages express implicit values and beliefs that form a coherent reality. When passing from one language to another one needs to experience the underlying values of these new realities, which involve our subjective experience that occurs through prereflexive and nonconscious processes. Thus, the act of translanguaging (*translangager*) includes out-of- and under-control skills, which operate back-and-forth continuously when students can combine lived emotional experience and reflective experience. It allow them to access an increasingly fine level of awareness of intra- and inter- subjective processes (Eschenauer, 2017: 141-142). In short, while for us the verb *translangager* refers to 'the dynamic act of relating to one's self, to others and to the environment through which shared meaning is constantly created between humans' (Aden, 2013: 115), the concept of *translangageance* describes the 'process of emergence of a common language that makes sense, through all forms of language' (Eschenauer, 2017: 15).

Empathy and aesthetic experience in a relational epistemology

The epistemology of enaction is relational as 'we work out our lives in a mutual linguistic coupling [...] because we are constituted in language (*langage*) in a continuous becoming that we bring forth with others' (Maturana & Varela, 1987: 234–235). The large body of research on affect, attunement, resonance and imitation in child-mother relationships confirms that language develops with empathy through affection and tenderness. Varela stresses this direct link between affection and empathy. For him, affect is preverbal and preconscious 'in the sense that I am affected or moved before an "I"– whoever it is – knows it'. He refers to 'the double dimension of the body (organic and phenomenal: *Körper* and *Leib*) which is an integral part of empathy, the best route to socially-conscious life beyond mere interaction, as a fundamental intersubjectivity' (1999: 16). Thus, the *trans* of language (*langage*) constitutes an interface between intersubjectivity and subjectivity and allows the construction of an individual's identity in connection with multiple social identities. For H.J. Krumm, 'language acquisition cannot be analysed as a cognitive process alone, but is always embedded in concrete social, historical, and individual biographical situations and is heavily emotionally charged. The individual and emotional initiatives of people interact with structures of society which either support or hinder such initiatives' (Krumm, 2013: 167).

In order to address the three-dimensional relational epistemology (self/others/environment) in language education, we suggest anchoring language learning in preverbal, phenomenal, experiential enactments of lived experiences. This is the very heart of Jacques Lecoq's pedagogy, which explores interactions before words through the *Leib-Körper* that knows things before we are aware of them (Lecoq, 1997: 41). Lecoq would ask his students to mime things and people in order to access the internal dynamics of meaning. According to him, experiencing the aesthetic dimension of relationships gives us access to a phenomenal knowledge of the world we share. In the international, multilingual school he founded in 1956, he developed an approach he called 'the body of words' based on his observation that the rhythm and dynamic force of words are a gateway to very subtle levels of understanding of language. He was thus convinced that the approach had enormous potential for language learning (1997: 61). Drawing from his teaching, we devise learning situations that enable students to perform creatively meaningful situations, at first nonverbally and then in their languages (*langues*), including the ones they are learning. Through performance, students are both actors and authors of their new knowledge and competences insofar as they embody the meaningful pieces of information emerging in the context.

An enactive pedagogy through a performative approach

Performativity has been considered across a range of academic areas. It was first introduced into the field of language philosophy by Austin (1962) and Searle (1969), who investigated the performative function of language and communication. Goffman (1974) analysed ordinary interactions as if they were performances, the social staging of oneself, and Butler (1988) used the term for the first time in cultural theory to mean we *are socially* what we do. Today, educators are exploring different models. Within the area of language teaching and literature studies Schewe (2011) advocates a 'performative didactic of foreign languages' that includes aesthetics through a drama-based teaching approach. It can be applied to the 'three core areas of a foreign language discipline (i.e. language, literature and culture)' (2011: 16). Within the science of education and performance studies, Sting (2012) relies on the concept of performativity in a transdisciplinary pedagogy. Beyond drama, he favours a phenomenological approach to education. According to him, the active, (inter)subjective, emotional and sensory nature of aesthetic experience can help reduce the gap between the unknown and the familiar. We ourselves think performance provides a relational matrix, a backdrop for the emergence of meaningful language in transdisciplinary experiences of learning that involve both situated action and dialogue.

Epistemological Framework of our Approach: A Cross Between Enaction and Aesthetics

Our research highlights the encounter of various fields of study that converge towards an epistemology of the Relation, notably linguistic performativity (Austin & Searle), sociolinguistics (Goffman) and bio-phenomenology from an enactive perspective (Varela).

In this section, we give a description of our enactive-performative approach based on key concepts of performativity defined by Fischer-Lichte (2004), who herself drew upon Varela's paradigm. Around the central concept of emergence, she highlights: experiential reality, the corporeality of language and thought and the transformative force of performance.

Table 5.1 Enactive-performative approach

	Enaction (Varela)	Performativity (Fischer-Lichte)
emergence	All forms of knowledge emerge in a structural coupling (perception/action). Perception is not something that happens to us but something we do/perform.	Performing is experiencing new realities through self-referential feedback loops between action and perception. Shared meaning emerges through performance. It is subject to the contingency of the context.
experience	'We work out our lives in a mutual linguistic coupling (...) because we are constituted in language in a continuous becoming that we bring forth with others' (Maturana & Varela, 1987: 234–235).	Performances do not seek to be understood but experienced. (Fischer-Lichte, 2004: 158)
corporeality	Knowledge is not something we acquire but embody through our sensations, emotions and movements that serve as the foundation of abstract thinking.	We act through our phenomenal body (*leib*) which conveys our consciousness of being alive and our semiotic or physical body (*Körper*). Both form our embodied mind. (Fischer-Lichte, 2004: 140)
transformation	Learning is a transformative process, 'an ongoing transformation in the becoming of the linguistic world that we build with other human beings' (Maturana & Varela, 1987: 235)	The transformative force of performance happens through three steps: 1) changing perspectives that estrange the familiar; 2) accepting a state of transition or openness that leaves room for the unknown; 3) embodying the new in the familiar and building a new reality.

How can this enactive-performative approach be implemented in language teaching? The context in which nonverbal and verbal language is produced is the teacher's primary concern, even before language correctness. The students are not advised to copy native speaker models but instead are encouraged to *live* and play with several languages, including the languages of their family and school. Movements, gestures, imitations are called into play to facilitate shared emergent meaning. When they become used to translanguaging (using all available verbal, emotional and kinesthetic repertoires), a balance is achieved between the body and the mind, which are biologically inseparable but often treated as dislocated in formal schooling. The didactic scenarios we devise are not written in advance by teachers outside the classroom. We set learning objectives (academic, methodologic and social/emotional aspects of learning [SEAL]) but we never plan students' linguistic and artistic productions beforehand so as to leave room for students' improvisation and creativity. The texts used in the performances are created by the students in several languages, sometimes supported by literary contributions or other artistic genres. These texts reflect their daily preoccupations while also linking with elements of the curriculum.

In classroom procedures relational objectives are brought to the fore through performance: students take on different roles, enabling the development of new perspectives. When they perform in new languages, they experience a paradox: the unknown becomes familiar and the familiar becomes strange. Aesthetic experiences thus develop emotional empathy which in turn develops flexibility in translanguaging.

The diversity of student identities provides the living material from which artists, teachers and students co-construct pedagogical scenarios, in class or in artistic workshops. By putting themselves in each other's shoes and using the languages of others, they transform their perception of reality. Our experiments have shown that the aesthetic eye the artists cast on the world, the students, and their languages has the power to move and transform students: their points of reference change and they see themselves and their transcultural identity in a new and compassionate way.

We see our pedagogical projects as *living enquiries* (Irwin *et al.*, 2018) conducted in communities of practice and research: artists, teachers, researchers and cultural mediators together creating projects that respond to the needs of the students. This helps us avoid the exploitation of Art as simply a means to an academic end, and instead preserves its key emergent, contingent, aesthetic and phenomenological nature. Although we take into account the content of school programmes, we invite students to question and make choices and decisions with adults in order to develop both their agency and sense of self-efficacy. Therefore, we advocate a practical experience of the arts, and notably drama, as a *polyaesthetic art* (Sting, 2008), through which students can feel things differently, change perspective and share original ideas. We also insist they attend live professional

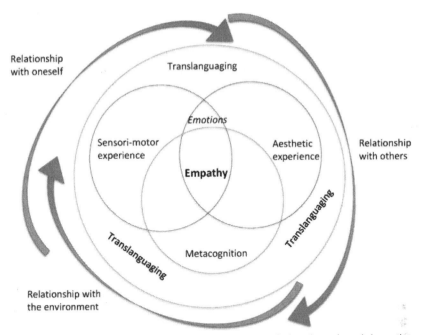

Figure 5.1 Mechanisms core to language learning revealed and employed through performative theatre (Eschenauer, 2017: 176)

performances in different languages to feed their imagination, and we facilitate meetings and exchanges with artists. These experiences echo what is happening in class and trigger an awareness of aesthetic dimensions of knowledge.

Our studies (Eschenauer, 2017, 2018b) reveal a balanced interdependance between the students' ability to translanguage, manage their emotions, and empathize. As shown in Figure 5.1, the more the students empathize, the more able they become to translanguage and *vice versa*. Aesthetic experience fosters empathy. Empathy is the driving force of flexibility which develops through translanguaging. All these phenomena emerge in a relational matrix (self/other/context).

An Example of an Enactive-Performative Project in a Plurilingual Environment

AiLES (Arts in Language Education for an Empathic Society) is a longitudinal study that took place in a pluricultural lower secondary school in a disadvantaged suburb of Paris. Sandrine Eschenauer followed a group of 20 students, aged 11 to 14 over four years (2011– 2015) in a bilingual class (where two foreign languages were taught). Although more than 16 languages were spoken and/or understood in this class, students felt embarrassed about, or even ashamed of, speaking their family languages. This attitude is reminiscent of:

> ... the centralized and standardized French educational system (in which) the French language served as an instrument of social and national integration and (in which) linguistic and cultural diversity was viewed as anathema to French public education (...) Even today, the notion of cultural "diversity" is not viewed favorably by French educators. (Kramsch & Aden, 2013: 49)

This is why languages of families are generally excluded in the school context and parents are sometimes advised to speak only French to their children to help them integrate

The study sheds light on the core mechanisms of empathy fostered by performative practices which promote the transition from one language and culture to another. It also shows that the capacity to translanguage helps students to develop empathic attitudes. Moreover it describes how aesthetic experience in all of the learners' languages directly correlates with the development of altruistic empathic abilities (Eschenauer, 2017: 490).

Artistic translanguaging workshops in foreign language teaching

We sought to restore these languages to their proper place and interweave a maximum of connections between them and the school culture by developing transdisciplinary projects aimed to link school subjects with artistic practices, school with families and teachers with artists in a holistic approach that gives a voice to the students in a meaningful learning context. The consequent experiences were meant to foster student creativity which, in turn, sought to enrich translanguaging (Aden, 2016; Eschenauer, 2018a). Each year, 20 artistic workshops of two hours were held, jointly or separately, by two actresses (German and English speakers). These workshops occurred during language classes and teachers also took part. Other artists (a photographer, a filmmaker and a dancer), who were gradually integrated into the project and all those who joined the group even for a short period of time (cultural mediators, trainee teachers, a philosophy teacher, an ethnologist, etc.), spoke exclusively in their mother tongues. The workshops formed the first spaces of translanguaging since we encouraged the students to use their family or home languages freely. When they chose to do so, they sometimes needed the help of their parents. They incorporated languages as a means of communication or as material for the artistic projects. They would use them in warm-ups and improvisations on the themes they had chosen, sometimes working from their language biographies and sometimes drawing inspiration from theatrical works they had seen. At the end of each year, they performed a piece in front of an audience and their parents. The performances were based on topics they had chosen and which they had explored with the artists and their English, German, literature, maths, music and art teachers.

The artists were never given academic objectives. They did not adapt their speech nor its flow for the students. They used mechanisms of kinesthetic and emotional empathy such as gestures, mimicry and vocal variations. When some students encountered difficulty, their classmates or teachers instinctively used verbal and nonverbal mediation strategies to avoid breaks in the theatrical work, thus developing collaborative attitudes.

Implementing a translanguage pedagogy

The project was implemented through reciprocal relationships. The teachers integrated the theatre sessions into their pedagogical objectives and, in turn, the actors were given these objectives as possible inspiration for the theatre sessions. When possible, the language teachers (English, German and French) built classes together using language and cultural awareness activities to highlight the similarities and differences between languages. They studied material using bilingual versions or handled a subject with documents in the different languages. For example, one team-teaching session explored the theme of superstition. After studying an extract from *Tom Sawyer,* the students discussed the cultural significance of the black cat in the story as a symbol of bad luck. The debate began in English and German and was extended by an ethnographic investigation into bad omens found in their own family cultures. This idea was included in their end of year performance.

In classroom interactions, the students used languages in recognisably bilingual ways, i.e. they moved fluidly between languages without being overly-concerned with the distinction between them, a point also noticed by the adults on the project. The students' choice of languages appears to have been intuitive. Driven by the desire to share their findings, they were not focused on form or choice of words; instead their utterances were pre-consciously regulated. In the excerpts below, we see how the students easily navigate between languages. To understand instructions, they either use nonverbal language or associate verbal and nonverbal clues. We noticed that very quickly, they instinctively chose the first language of the person they were talking to (German for the German actress or teacher), they were not afraid of using the wrong language and, when they became mixed up between two languages, this did not hinder the activity.

Warm-up exercise:

The lesson has not started yet, the students are talking amongst themselves. Both actresses (German and English) ask them to form a circle.

English actress -   Stand in a circle

German actress -   Kommt bitte in den Kreis!

They are not listening. One of the students, Chelsea, calls the class to order:

Chelsea   (mixing languages) – This time! Jetzt!

(...)

The German actress explains and demonstrates: when she walks everybody must walk, when she stops, everybody must stop. There is no verbal signal.

Chelsea -   Ah, ok, il faut la regarder ! Schauen, wir
                 schauen!

                 (*Oh! Alright, we need to look at her! Look!
                 We have a look!*)

Dylan: -   Comment on fait pour savoir ? (*How do we
                 know what to do?*)

Chelsea: -   You feel it!

                 (Excerpt from log book, 22/12/2012)

This desire to learn and share languages and the level of investment, solidarity and creativity of the students intensified at the end of the third year when they were touched by the public's emotional reaction to their final performance. As a result, they expressed their desire to work on what they called a 'real' theatre piece during their final fourth year. So we invited the well-known contemporary German playwright and director Lutz Hübner, who does not speak French, to work with us. After an initial discussion with the class, Hübner co-wrote a German-English piece with the students based on questionnaires about the characters they wished to play as well as improvisation workshops. The piece they named *Simply the Best* gave them a voice with which to discuss life in their neighbourhood, their desires, their hopes, their passions and their rebellions. With the help of their parents, the students included their family languages in the play's script at key and emotionally-charged moments.

In the excerpt from the play below, produced in 2015, the students realise they are being manipulated by the organisers of a reality TV show, and decide to speak in their family languages to prevent the presenters from understanding them. The script was handed to the students in German/English and they rehearsed directly using their extensive linguistic repertoires without writing their cues. They switched their parts a number of times, and they moved easily between languages accordingly. In the transcript, words uttered in family languages are underlined and transliterated where a non-latin script is used, followed by an English translation in italics:

ARNO (Presenter 1): Number two: Du da hinten. Wie heißt du? *(Hey you behind! What's your name?)*

Student 1 (user of Martinique Creole): Si ou pa ka songé non-nou, sé zafè kò ú! *(If you can't remember our names, that's your problem!)*

ISABELLE (Presenter 2): German or English!

Student 2 (Moroccan Arabic):

انتم ما تحترموش القواعد ، علاش تحنّوا احن نحترمها ntouma makathtarmouch dakchi, lach hna ghadi nhtarmouhoum? *(You don't play by the rules. Why should we?)*

ARNO: Wir verstehen dich nicht. *(We don't understand what you are saying!)*

Student 3 (Mandarin Chinese): 对啊，这才是你们的问题。Dui a, zhe cai shi ni men de wen ti. *(THIS is the problem!)*

ISABELLE: What?

Student 4 (Sinhalese): oba pavasana deya apa tērum nogatahot "banlieue" mata pradarśanayak tæbiya hækkē kesēda? *(How can you put on a show on the "banlieue" if you don't understand what we say?)*

(...)

ISABELLE (irritated): Will you start speaking German or English now!

Student 7 (Bambara): [inaudible] *(We don't feel like it!)*

The writer-director put the students' real-life strengths of plurilingual identity and translanguaging at the heart of the play. By the end of the experiment, the students no longer saw languages as a scholarly obligation disconnected from their interests.

Summary

Over the four years of the project, 86 workshops, 10 class-observations and 10 team-meeting sessions were filmed and/or audio recorded and subjected to thematic analysis. A selection of 12 semi-structured, 10 elicitation interviews and 66 qualitative questionnaires were cross-analysed. Each year, from 2012 to 2015, the students took the Davis psychometric test to assess any changes in their empathic attitude and all of them were interviewed on their updated language portraits. They also took a psychological test to assess their emotional state after 26 theatre workshops. The thematic content analysis revealed the following elements:

(a) Free from the pressure to be linguistically perfect, the students gradually became aware of their ability to translanguage. They developed great confidence in themselves, became more creative in their expression and sharpened their command of each language.
(b) Far from making it more difficult to learn, the simultaneous use of several languages in the artistic workshops connected the students to a rich emotional experience. They became less inhibited as the use of their other languages was never considered to be an obstacle or a mistake.
(c) The respectful, inquisitive and caring attitudes of the adults involved had a positive impact on the students' construction of identity and their sense of self-efficacy. Because each member of the project's community was recognised as an individual, the students better handled the linguistic and cultural diversity of their class.
(d) The main result of the study is the significant and balanced correlation between the development of empathy (emotional, kinaesthetic and cognitive), the intensification of aesthetic experience and the strengthening of translanguaging skills. The students made remarkable progress in their mental flexibility in language use and metacognitive activities.

Conclusion

How can education systems respond to the challenges of our century so that people of different habits, beliefs and languages can peacefully share an endangered and fragile ecosystem? Learning how to communicate is key to this, as is learning how to act together for a common purpose while fully respecting cultural diversity.

Transcultural education provides convincing responses for achieving such a goal as students are led to question the part they play in sustainable development by navigating between all languages and cultures, by taking on and respecting different points of view without giving up their own perspective. For Francisco Varela, 'intelligence can no longer be defined as the faculty to solve problems but as the ability to penetrate shared worlds' (1989: 113). The research we have presented is rooted in a relational paradigm that lies at the heart of translanguaging as a unique form of language practice (*translangeance*). This enactive turn has significant implications for language education. From a cognitive point of view, it means that we have to take into consideration advances in social and affective neuroscience in relation to the phenomenological dimension of perception in language teaching. From a sociological standpoint, we should go beyond advocating integrated teaching of all languages and instead seek social transformation so that diversity is recognised, experienced in transdisciplinary projects and valued as the bedrock of any education. The work of sociolinguists contributes to this humanist project, which seeks a new social balance that brings collaboration and altruistic empathy to the forefront. Clearly, the enaction paradigm linked to a performative approach to language education calls into question teaching methods constrained by standardised education programmes.

For our part, we propose to enact this humanistic vision in language education through a performative approach, as the aesthetic dimension of performance is crucial to restoring the balance of power between embodied and abstract knowledge. It allows the relational objectives to be implemented. Moreover, we believe that cooperation between artists, teachers, educators and researchers has the capacity to rehabilitate the sensory and emotional dimensions of languages that are all too often overlooked in teaching practices. We believe translanguaging (*translangageance*) can break down many barriers within educational systems still rooted in the primacy of reason over emotion.

References

Aden, J. (2008) *Apprentissage des Langues et Pratiques Artistiques*. Paris: Le Manuscrit.

Aden, J. (2010) L'empathie, socle de la reliance en didactique des langues-cultures. In J. Aden, T. Grimshaw, H. Penz (eds) *Teaching Language and Culture in an Era of Complexity: Interdisciplinary Approaches for an Interrelated World* (pp. 23–44). Brussels: Peter Lang.

Aden, J. (2012) La médiation linguistique au fondement du sens partagé: vers un paradigme de l'énaction en didactique des langues. In J. Aden and D. Weissmann (eds) *La Médiation Linguistique: Entre traduction et enseignement des langues vivantes. ELA* 167 (pp. 267–284). Paris: Klincksieck.

Aden, J. (2013) Apprendre les langues par corps. In Y. Abdelkader, S. Bazile and O. Fertat (eds) *Pour un Théâtre-Monde. Plurilinguisme, interculturalité et transmission* (pp. 109–123). Bordeaux: Presses Universitaires de Bordeaux.

Aden, J. (2016) Créer, innover par le théâtre: pour une pédagogie énactive des langues. Chapter 6. In I. Puozzo (ed.) *La Créativité en Éducation et en Formation. Perspectives théoriques et pratiques* (pp. 105–117). Bruxelles: De Boeck.

Aden, J. (2017a) Langues et langage dans un paradigme enactif. *Recherches en Didactique des Langues et des Cultures* 14 (1). http://rdlc.revues.org/1085 (accessed January 2020)

Aden, J. (2017b) Developing empathy through theatre: a transcultural perspective in second language education. In J. Crutchfield and M. Schewe (eds) *Going Performative in Intercultural Education: International Contexts, Theoretical perspectives and Models of Practice* (pp. 59–81). Bristol: Multilingual Matters.

Aden, J. and Pavlovskaja, M. (2018) Translanguaging and language creativity in drama staging. In G. Mazzaferro (ed.) *Translanguaging as Everyday Practice* (pp. 215–232). Cham: Springer.

Aden, J. and Eschenauer, S. (2014) Théâtre et empathie en classe bilangue: didactiser l'émergence. In N. Spanghero-Gaillard and E. Garnier (eds) *Pratiques Théâtrales en Classes de Langues. Les Langues Modernes* 4 (pp. 69–77). Paris: APLV.

Austin, J.L. (1962) *How to do Things with Words*. Oxford: Oxford University Press.

Bialystok, E., Craik, F. and Luk, G. (2012) Bilingualism: Consequences for mind and brain. *Trends in Cognitive Sciences* 16 (4), 240–250.

Butler, J. (1988) Performative acts and gender constitution: An essay in phenomenology and feminist theory. *Theatre Journal* 40 (4), 519–531.

Canagarajah, S. (2011) Codemeshing in academic writing: Identifying teachable strategies of translanguaging. *The Modern Language Jounal* 95, 401–417.

Creese, A. and Blackledge, A. (2010) Translanguaging in the bilingual classroom: A pedagogy for learning and teaching? *The Modern Language Journal* 94, 103–115.

Eschenauer, S. (2014) Combining foreign, native and national languages through theatre in schools: an "enactive approach" to education. In C. Nofri and M. Stracci (eds) *Performing Arts in Language Learning* (pp. 65–71). Rome: Edizioni Novacultur Rome.

Eschenauer, S. (2017) Language Mediation in Enactive Teaching at Secondary School. Longitudinal study of the connections between the phenomena of translanguaging, empath and aesthetic experience and their cognitive impact in the performative teaching of modern languages. PhD thesis in Language Science, Paris-Est-Creteil University, France.

Eschenauer, S. (2018a) Translanguaging and empathy: Effects of performative approach to language learning. In M. Fleiner and O. Mentz (eds) *The Arts in Language Teaching. International Perspectives: Performative – Aesthetic – Transversal – Europa lernen. Perspektiven für eine Didaktik europäischer Kulturstudien* (pp. 231–260). Berlin *et al.*: LIT Verlag.

Eschenauer, S. (2018b) Theatermethoden im Fremdsprachenunterricht. Ein Praxis – und Forschungsbericht aus einer Grundschule eines Pariser Vororts. In S. Asmus and V. Thielicke (eds) *Theatrale Methoden in anderen Fächern. Schultheater* 3 (4). Seelze: Friedrich Verlag

Fischer-Lichte, E. (2004) Ästhetik des Performativen. Frankfurt-am-Main: Fischer-Verlag.

Fischer-Lichte, E. (2012) *Performativität. Eine Einführung.* Bielefeld: Transkript.

García, O. (2009) Education, multilingualism and translanguaging in the 21st century. In A. Mohanty, M. Panda, R. Phillipson and T. Skutnabb-Kangas (eds) *Multilingual Education for Social Justice: Globalising the Local* (pp. 128-145). New Delhi: Orient Blackswan (former Orient Longman).

García, O. (2012) Theorizing translanguaging for educators. In C. Celic and K. Seltzer (eds) *Translanguaging: A CUNY-NYSIEB Guide for Educators* (1-6). New York: CUNY-NYSIEB.

García, O. and Li, W. (2014) *Translanguaging. Language, Bilingualism and Education,* London: Palgrave Macmillan.

Goffman, E. (1974) *Les Rites d'Interaction.* Lonrai: Éditions de Minuit, Coll. Le Sens Commun.

Husserl, E. (1921–1928/1973) *Zur Phänomenologie der Intersubjektivität. Texte aus dem Nachlaß, 2. Teil: 1921–1928.* The Hague: Iso Kern.

Immordino-Yang, M. H. (2015) *Emotions, Learning, and the Brain. Exploring the Educational Implications of Affective Neuroscience.* New York/London: W.W. Norton and Company.

Irwin, R., LeBlanc, N., Ryu, J. and Belliveau, G. (2018) A/r/tography as living inquiry. In P. Leavy (ed.) *Handbook of Arts-Based Research* (pp. 37–53). New York/London: The Guilford Press.

Kramsch, C. and Aden, J. (2013) ELT and intercultural/transcultural learning. An overview. In J. Aden, F. Haramboure, C. Hoybel and A.-M. Voise (eds) *Approche Culturelle en Didactique des Langues* (pp. 39–59). Paris: Éditions le Manuscrit.

Krumm, H.J. (2013) Multilingualism and identity. What linguistic biographies of migrants can tell us. In P. Siemund, I. Gogolin, M.E. Schulz and J. Davydova (eds)

Multilingualism and Language Diversity in Urban Areas: Acquisition, Identities, Space, Education (pp. 165–176). Amsterdam: John Benjamins Publishing Company.

Lachaux, J.P. (2011) *Le Cerveau Attentif. Contrôle, Maîtrise et Lâcher-Prise*. Paris: Odile Jacob.

Lecoq, J. (1997) *Le Corps Poétique: Un Enseignement de la Création Théâtrale*. Paris: Actes sud-Papiers.

Maturana, U. and Varela, F. (1987) *The Tree of Knowledge: The Biological Roots of Human Understanding*. Boston: Shambhala.

Merleau-Ponty, M. (1945) *Phénoménologie de la Perception*. Paris: Gallimard.

Schechner, R. (2002) *Performance Studies*. London/New York: Routledge.

Schaeffer, J.M. (2015) *L'Expérience Esthétique*. Paris: Gallimard Essais.

Schewe, M. (2011) Die Welt auch im fremdsprachlichen Unterricht immer wieder neu verzaubern. In A. Küppers, T. Schmidt and M. Walter (eds) *Inszenierungen im Fremdsprachenunterricht* (pp. 20–31). Kempten: Diesterweg.

Searle, J. (1969/1972) *Les Actes de Langage*. Paris: Hermann.

Sting, W. (2008) Anderes sehen. Interkulturelles theater und theaterpädagogik. In K. Hoffmann and R. Klose (eds) *Theater Interkulturell: Theaterarbeit mit Kindern und Jugendlichen* (pp. 101–108). Berlin: Schibri-Verlag.

Sting, W. (2012) Performative praxen und theaterpädagogik. In W. Sting, G. Mieruch, E.M. Stüting and A.K. Klinge (eds) *Poetiken des Theatermachens, Werkbuch für Theater und Schule* (pp. 119–126). Bobingen: Kessler Druck und Medien Kopaed.

Thirioux, B., Mercier, M.R., Blanke, O. and Berthoz, A. (2014) The cognitive and neural time course of empathy and sympathy: An electrical neuroimaging study of self-other interaction. *Neuroscience* 267, 286–306.

Varela, F. (1989) *Invitation aux Sciences Cognitives*. Paris: Seuil.

Varela, F. (1996) Neurophenomenology. A methodological remedy for the hard problem. *Journal of Consciousness Studies* 3 (4), 330–349.

Varela, F. (1999) Quatre phares pour l'avenir des sciences cognitives. *Revue Théorie Littérature Enseignement* 17, 7–21.

Varela, F. (2002) Autopoïese et émergence. In R. Benkinrane (ed.) *La Complexité, Vertiges et Promesses. 18 Histoires de Science* (pp. 161–176). Paris: Editions Le Pommier.

Varela, F. (2017) *Le Cercle Créateur. Écrits (1976–2001)*, ed. M. Bitbol with A. Cohen-Varela, J.-P. Dupuy and J. Petitot. Paris: Seuil.

Varela, F., Thompson, E. and Rosch, E. (1993) *L'Inscription Corporelle de l'Esprit: Sciences Cognitives et Expérience Humaine*. Paris: Seuil.

6 Theorising Arts-Based Collaborative Research Processes

Jane Andrews, Richard Fay, Katja Frimberger, Gameli Tordzro and Tawona Sitholé

Introduction

In its report into interdisciplinary research entitled *Crossing Paths* (2015) the British Academy elaborated on what they identified as the potential benefits, but also the many challenges, of working across disciplines to achieve research goals which address enduring problems in the world. In this chapter we explore some of the issues raised in the British Academy report as we believe they resonate with our own experiences of how creative arts methods can be incorporated into an interdisciplinary research project. We document and analyse how we used arts-based methods from the outset in our work, which was built around a collaboration across disciplines, modes and professions on the AHRC funded project *Researching Multilingually at the Borders of Language, Law, the Body and the State*, (AH/L006936/1). We theorise here our collaborative and transformative practice by drawing upon thinking firstly from new materialists (e.g. Barad, 2003; Dolphijn & van der Tuin, 2012) and secondly from *interthinking*, a concept developed by applied linguists and educational psychologists (e.g. Littleton & Mercer 2013). New materialism (discussed in detail later in this chapter) explores what Barad (2003) names *intra-action* between components of phenomena which shape each other by coming into contact with each other. For our work this could involve examples of discussion of research planning being shaped by, and transformed by, exploration using metaphor and dramatisation. Interthinking (also discussed later in the chapter) supports our understanding of how the different modes of communication, including languages, facilitate collaborative problem-solving in teams. Translingual practice (drawing upon Canagarajah's 2013 term for the flexible uses of language and other modes of communication for specific effects whether communicative, poetic, activist or other) occurred regularly in our collaborative work reported in this chapter (see also Fay *et al.*, 2016, for deeper discussion of translingual practice).

The chapter takes the following structure: (i) background to the interdisciplinary research project; (ii) selected principles from new materialist thinkers and writers on interthinking in language and education, as deemed relevant to this chapter; (iii) the presentation and discussion of two vignettes from the project of arts-based, collaborative, transformative and translingual practice; and, (iv) a conclusion with reflections on implications for future research projects seeking to implement and conceptualise their own interdisciplinary, arts-based research.

The Researching Multilingually at Borders Project

Between 2014 and 2017, our international team of creative artists and academics covering areas including music, theatre, poetry, textiles, law, global mental health, modern languages, applied linguistics, anthropology, sociology, intercultural communication and education collaborated on a research project exploring the role of language in contexts of pain and pressure. The sites for the research were defined in a wide range of ways and included, in brief, an ethnography of border processes in Bulgaria and Romania (see Gibb & Danero Iglesias, 2017), an online community of learners of Arabic initiated by the Islamic University Gaza (see Imperiale *et al.*, 2017), the learning of modern languages in the border state of Arizona, USA (see Gramling & Warner, 2016), the legal practice spaces in the Netherlands and Scotland in which asylum appeals are heard (see Zwaan, 2010; Craig, 2014), a global mental health intervention in Uganda (see White *et al.*, 2015) and the further education classroom in Scotland where young people who are unaccompanied refugees meet and learn together (see Frimberger *et al.*, 2017). These sites illustrate the diverse and interdisciplinary ambition of the research project and provided the opportunity to explore potential challenges, as noted in the British Academy (2015) report, that such work across disciplines often entails. The case studies also serve to contextualise the people and sites to be engaged in the arts-based research processes of the project.

The project had two overarching aims which were:

(1) to research interpreting, translation and multilingual practices in challenging contexts, and,
(2) while doing so, to evaluate appropriate research methods (traditional and arts based) and develop theoretical approaches for this type of academic exploration.

In practice, the academic disciplines represented in the project operated as five case studies, each of which was guided by its own research questions. There were also two hubs (covering creative arts and researching multilingually, henceforth CA and RM) which worked collaboratively together and across the case studies and similarly were guided by their own research questions. The authors of this chapter were members of the two hubs (Frimberger, Sitholé and Tordzro were members of the CA hub and Andrews and Fay were members of the RM hub). The RM hub had the following as their guiding research questions:

(a) How do researchers generate, translate, interpret and write up data (dialogic, mediated, textual, performance) from one language to another?
(b) What ethical issues emerge in the planning and execution of data collection and representation (textual, visual, performance) where multiple languages are present?
(c) What methods and techniques improve processes of researching multilingually?
(d) How does multimodality (e.g. visual methods, 'storying', performance) complement and facilitate multilingual research praxis?
(e) How can researchers develop clear multilingual research practices and yet also be open to emergent research design?

The CA hub had the following statement as its guiding, overarching objective:

> to produce creative synthesis and a constant dynamic of translation between languages, place and media throughout the project.

This will be achieved by

(1) Translation of research data, concepts and findings from academic form into live performance, through the creation of a playtext by the project playwright, Tawona Sitholé, based on data from all the project components, which will be rehearsed, produced and performed in Scotland, Ghana, and other countries in which research will be conducted;
(2) Community drama and rehearsals in each of the case study sites where forum theatre workshops will allow for the exposure of otherwise silent dynamics of language, power, narrative and pain relevant to participants in each context and the RMTC 'hub';
(3) Capacity building and training across the project and its different contexts in using performance to represent specific translation and interpretation processes and practices (and issues such as silence and 'the untranslatable') through other media.

The project made use of alternative conceptualisations of linguistic competence in real world contexts, namely, Phipps' (2013) linguistic incompetence and the use of arts methods as languages in their own right, with equal status to oral and written languages. In her 2013 paper, Phipps explored how, by reflecting on everyday realities of migration in the 21st century, it may no longer be of use or value for researchers and practitioners to maintain a belief in the attainment of linguistic competence as the desirable outcome of language learning which necessarily precedes engaging with users of a specified language. Phipps' alternative way of thinking, based on her own experience, is to appreciate the potential of speakers who are in more socially powerful positions to express their linguistic solidarity by showing their linguistic *incompetence* when using a language that is not expected to be used in a particular context. Phipps advocates the benefits of being a beginner learner of a language and showing oneself to be a learner of a language when interacting with people who may be experiencing pressure and pain, for example in the context of seeking asylum.

This conceptualisation of language provided a foundation for our researcher team to embark on our research (with each other and with our collaborators and research participants) with an awareness of how, in real life contexts, expressions of solidarity may come from language incompetence as well as language competence. It also opened up the opportunity for the research team to name and acknowledge the linguistic resources held between us (they were: Twi, Tigrinya, German, Italian, Arabic, French, English, Dutch, Bulgarian, Turkish, Danish, Shona, Ndau). The use of these resources in our research practice, in spite of, and because of, our varying levels of competence, is illustrated and discussed further below.

In addition, Phipps' work (e.g. 2013), and that of Frimberger *et al.* (2017), have focused on the use of the arts as a language which, they argue, has great potential to communicate through diverse modes (visual, oral, written) in contexts where participants have experienced trauma. In our project we collaborated in using these specific arts-based techniques throughout the duration of the project in our research practice: (i) working with printing and textiles using symbols valued in specific cultural contexts (e.g. see Tordzro, 2016, for research practice using Adinkra symbols); (ii) poetry and spoken word (Phipps *et al.*, 2016); (iii) craft techniques (Frimberger *et al.*, 2017); (iv) dance; and (v) drama. The discussion in this chapter is based, as noted above, around two vignettes of researcher-artist collaboration within the research team. Other examples of arts-based practice which involved research participants beyond the research team (e.g. clients in an NGO for refugees in Romania) also took place but these are not the focus of this chapter.

Theorising Arts-Based, Translingual, Multimodal, Collaborative Research Processes

New materialism is: 'A *performative* understanding of discursive practices [which] challenges the representationalist belief in the power of words to represent pre-existing things' (Barad, 2003: 802). Barad continues by stating that performativity is 'a contestation of the excessive power granted to language to determine what is real'. This way of explaining how we construct and represent reality encourages us to move away from a reliance on language as the main medium and invites attention to modalities such as, in the case of our research, music, visual arts, dance, theatre and poetry. Barad's challenge to an over emphasis on the power of words also leads her to address the dangers inherent in habitualised, disciplinary-bound ways of thinking and researching and as such, has particular relevance for our interdisciplinary and arts-based project. She states (2003: 810) that

> If we follow disciplinary habits of tracing disciplinary-defined causes through to the corresponding disciplinary-defined effects, we will miss all the crucial intra-actions among these forces that fly in the face of any specific set of disciplinary concerns.

The concept of intra-actions invites us to pause on and notice what happens when experiences, ideas and phenomena meet and have an impact on each other. These intra-actions, for our project, could have taken multiple forms, such as, uses of arts-based methods in contexts where they might not conventionally be practised, using methodologies within disciplines which are not usually associated together or prioritising certain languages in research and de-prioritising others (e.g. taking an English-last approach, Phipps, 2017). For Barad (2003: 802) 'the move towards performative alternatives to representationalism shifts the focus from questions of correspondence between descriptions and reality (e.g. do they mirror nature of culture?) to matters of practices/doings/actions.' This emphasis on the importance of doings and actions, along with the call to reject a reliance on language, can provide a new materialist theoretical framing for our arts-based, interdisciplinary research work which sought to explore human experiences in contexts of pressure and pain and to remain aware of how languages or silence were used in such encounters. The vignettes provided below are offered as examples of our work and can be considered as doings and actions.

In her doctoral research exploring migratory aesthetics in what she names arts-based strangeness research with international students in Glasgow, Frimberger (2016) makes use of new materialist thinking. Frimberger draws upon Patti Lather's work (2013) in analysing images generated in arts workshops in which she explores knowledge production which is not limited by linguistic description but rather, 'language, bodies, objects and the environment are seen as involved in a constant, meaning-making process' (Frimberger, 2016: 5). Frimberger's contribution to this chapter and the research project, along with the CA hub colleagues, was to encourage doings, actions and intra-actions which moved the research team beyond languages-based explorations of our project research questions. However, the commitment to integrating arts-based practices into the project from its earliest days and the influence of new materialist thinking did not preclude our attending to how our project-wide linguistic resources were used in our collaborative working. The next section focuses on how Littleton and Mercer's (2013) work on interthinking shaped our reflections on how interdisciplinary discussion was also central to our research.

Interthinking

There is a large body of work in the education literature which analyses and promotes dialogues as a foundation of learning (see, as examples, Alexander, 2017; Brice Heath, 1983; Littleton & Mercer, 2013; Wells, 1986). These works have focused on peers (adults or children) interacting together in specific contexts such as school classrooms and university seminar rooms. They have also explored adult-child interaction in homes and teacher-pupil, lecturer-student interactions in educational settings. In such studies, researchers have focused on a wide range of linguistic phenomena such as aspects of language acquisition and ways in which parents scaffold children's utterances (e.g. Wells, 1986) and cultural patterns of interaction in communities and families (Heath, 1983). Alexander (2017) takes as his starting point the educational benefits of dialogic approaches to teaching and learning, and not the uses of and features of language in use in their own right. Littleton and Mercer (2013) also focus their work on what language is used for by groups of speakers and how it achieves, or not, those purposes. To delineate this specific research focus Littleton and Mercer coin the term *interthinking*, which they apply to discourse analyses of interactional data from group activities in real-life settings such as musicians in band practice sessions, workplace meetings between colleagues, and pupils interacting while using an interactive whiteboard. Within their data sets, the authors explore talk as (2013: 13) 'a social mode of thinking – language as a tool for teaching-and-learning, constructing knowledge, creating ideas, sharing understanding and tackling problems collaboratively'. We note that in the first example of this type of social thinking, which is facilitated through talk (interactions in band practice sessions), that language is not the sole mode of communication as music itself serves to generate ideas and initiate problem-solving.

A feature of the writing on interthinking from Mercer (e.g. 2000) is that not all dialogues are deemed to be effective in terms of achieving the intended collaborative goals. Three types of talk are defined: disputational, cumulative and exploratory, with exploratory talk, in which speakers build on, challenge or generate new thinking or actions, being designated as the most effective for achieving goals. In our large, interdisciplinary and arts-based research project we met together as a whole team of 20 people and as well in smaller case study plus hub teams. As has been noted in one of the few published explorations of team-based interactions in a large research study (see Creese & Blackledge, 2012), the work of meaning-making within research can be seen in the space of project meetings. As the authors observe:

> whole-team meetings provide a window into the process of analysis as colleagues brought to the table their emergent understandings of the phenomena under investigation. In these meetings we did more than listen, arguing, negotiating, contesting, agreeing, introducing our different perspectives and histories as we attempted to make meaning out of observed linguistic practices. (Creese & Blackledge, 2012: 307)

Creese and Blackledge (2012) articulate the complexity of their process of discussion as a research team, involving contestation as well as agreement, and this work can be seen as sharing features of Littleton and Mercer's (2013) interthinking. In the vignettes below we present extracts of collaborative practice involving interthinking which, in the context of our project, was also mediated by material means such as textiles as well as drawing upon the linguistic resources within the team to stimulate discussion, shared poetry writing and reading, and reflective writing.

Collaborative Transformations

In this section we offer two vignettes of what we name collaborative, translingual and transformative practice between the two research hubs in the RM project team. The doings and actions of these two vignettes took place within the first year of the three-year project and as such formed part of the exploratory work establishing how we could work together across modes, and professional and disciplinary borders. In each vignette some elements of the first-hand experience of collaborative working are offered, with an accompanying contextualisation of the work, and then a discussion follows which draws upon ideas from new materialism and interthinking, as introduced above.

Vignette 1: Working with Metaphor and Movement – The Well

Vignette 1 sets out an experience of the joint hub working on what we named hotspot texts. One year into the three-year project, a joint hub meeting was agreed in which ideas could be exchanged between the CA and RM hubs. In the spirit of working closely with creative arts methods throughout the project, the CA hub invited the RM hub to spend a day in their workspace which they named a lab, signalling the intentionally experimental nature of the work carried out there. To facilitate the exchange of ideas and collaboration between us all, Richard Fay proposed that we each write about one or more *hotspots*, ideas or experiences identified from the project so far as being thought-provoking or puzzling and which might merit creative exploration together. The term 'hotspot' was deliberately chosen as not being a typical term used within academic research or as belonging to a particular methodological tradition which might constrain the nature of our interaction. Rather, it was hoped that the hotspot concept would open out our discussion and collaboration and allow us all to embrace creative arts and multimodal dimensions.

In practice, each colleague wrote approximately one paragraph for each of their hotspots (typically between two and four per person). The hotspots covered areas ranging from nhorwa, tasaamuh, speech and writing, to researching interculturally. In keeping with the project's attention to researching multilingually, as seen in this list, colleagues drew on their linguistic and cultural resources to shape their reflections. Nhorwa was offered by Tawona Sitholé (CA hub) and it drew upon a word and concept from his linguistic resource of Shona and of a practice in which gifts were offered to guests. Tawona's hotspot was how our interdisciplinary and multimodal research and collaboration could involve us in sharing gifts of insights drawing on our distinctive sources of knowledge and experience. Tasaamuh was offered by Mariam Attia (RM hub), originated in Mariam's Arabic resources and referred to the concept of acceptance or tolerance or letting be, which Mariam explained as being relevant to her understanding of how to engage together in a complex, interdisciplinary, multi-site project.

The development of the hotspot texts took place with all researchers in their home university and workplace, all within the UK but not in the same cities. A desired next stage was to meet face-to-face and move into a creative consideration of the ideas built into the hotspots, a move agreed by all as essential to continued joint working. We saw this as an opportunity for the RM hub members to work in the lab, or creative space, of the CA team, in Glasgow in a venue which was not part of a university. Once in the lab space, Tawona asked for us all to propose one or more metaphors which emerged for each of us from our thinking about our own and each other's hotspots. One of the RM team members, Mariam Attia, proposed the metaphor of a well. Mariam explained for researchers to draw upon their linguistic resources in doing their research multilingually, there is an analogy with the act of drawing water from a well. For Mariam, the action is perhaps arduous but the rewards are plentiful. Guided by Tawona, an agreement was reached (quite quickly) to spend longer exploring the metaphor of the well and to move into different modes. To do so, all hub members worked in two groups and used movement, music and mime to explore the metaphor of the well. It was noted by us all that our individual and group work using movement brought us together to produce a collective output but was also shaped by our individual understandings of the actions and motions involved in drawing water from a well. These understandings, we reflected, were clearly shaped by our cultural and experiential lives so that novices and experts were apparent. We all felt that the metaphor of drawing water from a well opened up our understanding of working multilingually together in a research team. The well metaphor became, subsequently, a shared point of reference, which was returned to many times in future collaborative encounters. Tawona reflected on the collaborative learning of the two hubs after the movement and music work on the 'well' metaphor:

> The 'well' had such an impact because it was another creative moment from the so-called non-creative team; it validated the use of exploration as a method of working; it helped us unlock meanings; and it has a universal sense and appeal. (Tawona Sitholé, Reflections on Hotspots document, May 2015)

Future team discussion of the well metaphor served as a reminder of our dual sense of the toil and reward involved in accessing the resources of a well. The photo below was taken during a visit to a city where part of our second case study was taking place (Bucharest, Romania) and provided a visual reminder of our metaphor work.

By working together with metaphors the two research hubs engaged in interthinking to explore aspects of the project's objectives and methods as identified in the hotspots. The metaphor-work involved using words (generating ideas, proposing and accepting metaphors and reflecting back on processes and movement work) and also took us into embodied ways of collaborating where words were not prioritised or dominant. Writers such as Barad (2003) working in the new materialism paradigm deploy the notion of *entanglement* to convey how meanings and materials or matters operate together and have an impact on each other. Our physical work with our bodies to mime our engagement with a metaphorical idea which began with a reflection on our project's work throughout the free writing of

Figure 6.1 The well

hotspots can be considered as an act of entanglement, we argue here. We sought to identify, present and re-present shared ideas and understandings to each other in order to get to know them better and see how they, and we, intra-acted in our collaborative work together.

Vignette 2: Identifying Hotspots of Curiosity – Moving from Themes to Poetry

A second way in which the eight project members collaborated in the CA lab with their hotspots involved a series of activities, initiated by the CA hub and inspired by their creative ways of working informed by poetic, theatrical and musical techniques. The hotspots were performed with each one being read aloud by someone who was not the original author. This had the effect of presenting the hotspot for consideration by the team but in oral mode and with the inevitable lack of fluency which comes when readers are not familiar with the words or meanings lying behind what they are voicing. A reflective discussion of this process led to a proposal from the CA hub that a new stage of engaging with the hotspots would be to transform the hotspots into poetry. This stage would involve working with the original hotspot to select and re-present it in a poetic form. The process was engaged with in different ways by different team members as shown below (Figures 6.1 and 6.2). RM hub member Jane Andrews' hotspot was transformed into a poem by CA hub member Tawona Sitholé, a published poet, whereas RM hub member Richard Fay worked with his own hotspot to explore the process of transformation from hotspot to poem himself. This across-hub collaboration in the CA lab space allowed time and opportunities for exploring and being playful with language in our research process.

Jane's hotspot	(Jane's poem, authored by Tawona Sitholé)
Thinking about what can be achieved in speech and writing in both the project (e.g. the RM hub 'Ways of working' document) and in the research sites themselves. Initial examples include i) within contexts we have explored with a project doctoral student exploring legal processes and with Alison's (the project PI) guidance, and ii) within collaborative research work e.g. in conversation with	researchers and participants move between modes of communication as well as between languages benefits and disadvantages spoken and written texts and interactions shape the work being undertaken how do we generate ideas together and then capture them moving between spoken and written texts

Tawona he expressed his preference for discussing ideas, exploring 'burning issues' as more valuable than only exchanging written texts (emails or attached notes). This has encouraged me to think about the work of Karen Littleton and Neil Mercer (inspired by Vygotskian thinking) on what they call 'interthinking' and 'putting talk to work'.	or moving beyond them in creative ways

Richard's hotspots	**(Richard's self-authored poem)**
Multilingual Practice in Research Collaboration *I have been intrigued by our individual, shared, and collective practices within the project vis-à-vis the uses of a variety of languages, by our attempts to be linguistically inclusive and diverse and not to remain solely within the lingua franca English-medium academic discourse. There are lots of moments to reflect upon in this regard - our use of a word in any language at the end of the Glasgow meeting to capture our 'take' on those intense days of work together; Tawona's use of Shona during our events (e.g. 'Hekani'; his walk-about poem on the 2nd day of the Durham symposium); Robert and Julien's English/ French spoken script/Powerpoint combinations; Nazmi's use of English and Arabic in his Durham Skyped presentation/ ppt; and so on. For me this is a hotspot because my own reaction to these attempts is not yet clear.*	and my brother asks me, as brothers do, "so exactly what parts of your skill-set qualify you for the project in uganda?" hmmm . . . I'll start simply .... I play music, not really a creative artist, more a 'musiker' I play with languages, not really a linguist, more a 'languager' but it gets complicated ... I support others, not a therapist, just a ??? hmmm, 'well-beinger' I explore 'cultures', not an anthropologist, just an 'interculturalist' and even more complicated ... I fight injustice, not a lawyer, just an activist and I've never worked under siege, just in awe of what those who are can achieve I'm thinking about researching multilingually, but I am not researching multilingualism and it gets simpler ....

Nor is my response to the comments of others (e.g. Angela Creese's comments on Robert/Julien's English/French mix; Alison's current strategy of reading authors in languages other than English and/or authors writing in English as a second language, and her reminder of the political scholarly habit of some feminist researchers of reading only works by women). Thus, I am left with an ongoing puzzle: given that we want our (academic as well as research) practice to be purposeful, but also that we want to challenge, in our own fields, the linguistic hegemonies that frame experience more widely and which contribute to the borderlands where discrimination, inequity, disadvantage, voiceless-ness, and trauma can be created and compounded by language hierarchies, how we might position ourselves vis-à-vis the 'wider currency' that English (and other privileged languages) have, a currency that for me raises all my post-colonial and UK-based angst (about Englishes) but for others (e.g. Elena from Cordoba in the RM-ly Network project) can be more a matter of linguistic hospitality and/or a strategic choice to embrace the power that this language resource has acquired?

Figure 6.2

The poetry based on the original hotspots was collected into a playscript by CA hub member Tawona Sitholé and formed part of a performance (from both hubs) which was shared in a whole team meeting later in the academic year. The content of the hotspots is not the focus of this chapter, although it is worth noting briefly that the areas explored here by Jane and Richard dwell on issues such as, for Jane, how linguistic modes play out in collaborative research in terms of when and how speech and writing are used and, for Richard, how collaborative multilingual practice in research can be achieved. These areas for exploration indicate that, at the time of writing, the two writers did not feel a sense of certainty about how they should or could carry out their research but at the same time they felt they could identify areas which would merit further researcher attention.

The process of working with our hotspots and performing them in different ways (e.g. performing to those who had written the original versions; performing in new versions to the wider project team) had the effect of bringing us into close awareness of each other's observations and interpretations of phenomena on the project. In addition, the processes enabled us to spend extended time on working with ideas, words and sounds due to the stages of sharing and transforming them into different formats. This, we believe, supported our collective reaching of 'deeper meanings' as noted by Gameli Tordzro (CA hub member). The deeper meanings were enriched by poetic forms and by concepts and terms from several, not just one, language, keeping us close to our researching multilingually intentions in the project. Because we moved between free writing (of hotspots) to performed reading to poetry to performed script, we feel this was an emergent way of collaborating which sought to move far away from what might have been a conventional academic research process. New materialist thinking has encouraged us to define our process as valuing doings and actions over linguistically-constructed analyses of data. It has also allowed us to pause on and struggle with the complexity of what the project was concerned with and our research process. The outcome of it led us to identify future areas of exploration guided by what had been fruitful in our collaborative activities up to that point, thereby serving an agenda-setting function in our work.

The essence of the interthinking concept, explored earlier in this chapter, is that exploratory talk, or it might be multimodal dialogue, enriches human thinking and understanding through the collaborative engagement on problems or concerns held in common. In Vignette 2 the collaborative aspect was built into the process through, initially, the sharing of individual writings and, subsequently, the continued working with the ideas embedded in those writings. The writing became more creative as the collaboration developed (first poetry, then performance script) rather than remaining at a representational stage of research experience. As such our interthinking process was multimodally, and materially, shaped, thereby bringing together the two theorisations of interthinking and new materialism.

Concluding Reflections

In this chapter we have theorised and exemplified an approach to weaving creative arts methods into a complex, multi-site, multilingual and interdisciplinary research project. Our work with creative approaches from the beginning of the project supported a different way of engaging with ideas, with each other as team members, and with participants in the sites of the case studies. This different way allowed us to engage with the materiality of the phenomena we as researchers were encountering and it moved us beyond a linguistic representation of our researcher experience and the experiences of the participants in our case study research into a multimodal form. Our approach and our theorisation using new materialism also enabled us to consider how we intra-acted in the study and this, we hoped, steered us away from othering our experiences or the people we worked with.

Our large research team sought to learn from each other from different disciplines, different expertise and experience in creative arts, different linguistic resources, and as such our collective discussion displays many of the features of interthinking. According to Littleton and Mercer (2013) effective interthinking also involves co-production of new ideas and understandings and can promote the appropriation of ideas across the group. In the case of our engagements, the forms of appropriation could be visualised and metaphorical, as was shown in the use of the well metaphor and the subsequent within-team referencing of this shared understanding. These multimodal forms stayed with us and were available for us to build on as common tools for our continued communication.

The prioritising of creative arts methods continued throughout the life of the project and has expanded the range of project outputs into film e.g. Tordzro (2016) *The Calabash People*, and Frimberger and Bishopp (2014) *Welcome to Scotland*. Our groundwork in the first year of the project, which is reported in the two vignettes in this chapter, established for us our collective commitment to working collaboratively, across disciplines, art forms, and languages. We believe that for researchers who are also committed to interdisciplinary and creative approaches to their research, time spent identifying and grappling with core concepts in a variety of modes can provide an essential foundation for the future effective completion of the research. To return to the British Academy's (2015) report on interdisciplinarity in higher education, we can see parallels between that report's findings and our own experiences of working with creative methods in a large, interdisciplinary research project. In the words of the report where UK higher education and research has not embraced interdisciplinarity:

> It has perhaps nurtured a relatively simplistic approach in some disciplines, emphasising rigorous demonstration of cause and

effect at the expense of efforts to understand society as a complex system of physical, technological, environmental, social, economic, political and cultural processes and feedback loops. (British Academy, 2015: 22)

In our creative arts and interdisciplinary work we have grappled with entanglements and sought to reach deeper meanings at every stage of our collaborative process.

Acknowledgements

We are grateful to Mariam Attia for her generous collaboration on the project and her agreement to be referred to in this chapter.

References

Alexander, R. (2017) *Towards Dialogic Teaching: Rethinking Classroom Talk* (5th edn). York: Dialogos

Barad, K. (2003) Posthumanist performativity: Toward an understanding of how matter comes to matter. *Signs* 28 (3), 801–831.

Barad, K. (2007) *Meeting the Universe Halfway: Quantum Physics and the Entanglement of Matter and Meaning*. Durham and London: Duke University Press.

Brice Heath, S. (1983) *Ways with Words*. Cambridge: Cambridge University Press.

The British Academy (2015) *Crossing Paths: Interdisciplinary Institutions, Careers, Education and Applications*. London: The British Academy.

Canagarajah, S. (2013) *Translingual Practice: Global Englishes and Cosmopolitan Relations*. London and New York: Routledge.

Craig, S. (2014) Case comment: Secretary of State for the Home Dept v MN. *Journal of Immigration, Asylum and Nationality Law* 28 (3), 293–296.

Creese A. and Blackledge, A. (2012) Voice and meaning-making in team ethnography. *Anthropology and Education Quarterly* 43 (3), 306–324.

Dolphijn, R. and van der Tuin, I. (2012) *New Materialism: Interviews and Cartographies*. Ann Arbor: Open Humanities Press.

Fay, R., Andrews, J., Frimberger K. and Tordrzro, G. (2016) Creative Interthinking: Interthinking creatively presentation and performance. Paper presented at *12th International Congress of Qualitative Inquiry*, University of Illinois at Urbana-Champaign, 21 May 2016.

Frimberger, K. (2016) 'Struggling with the word strange my hands have been burned many times': Mapping a migratory aesthetic in arts-based strangeness research. *Studies in Theatre and Performance* 38 (1), 9–22.

Frimberger, K. and Bishopp, S. (2014) *Welcome to Scotland*. www.youtube.com/watch?v=bl77CaxL_BA (accessed January 2020)

Frimberger, K., White, R. and Ma, L. (2017) 'If I didn't know you what would you want me to see?': Poetic mappings in neo-materialist research with young asylum seekers and refugees. *Applied Linguistics Review*: Special issue on Visual Methods 9 (2–3), 391–419.

Gibb, R. and Danero Iglesias, J. (2017) Breaking the silence (again): On language learning and levels of fluency in ethnographic research. *Sociological Review* 65 (1), 134–149.

Gramling, D. and Warner, C. (2016) Whose 'Crisis in Language'? How translingual students critically reframe the future of foreign language learning. *L2 Journal* 8 (4), 76–99.

Imperiale M.G., Phipps A., Al-Masri N. and Fassetta G. (2017) Pedagogies of hope and resistance: English language education in the context of the Gaza Strip, Palestine. In E.J. Erling (ed.) *English Across the Fracture Lines: The Contribution and Relevance of English to Security, Stability and Peace* (pp. 31–38). London: British Council.

Lather, P. and St. Pierre, E.A (2013) Post-qualitative research. *International Journal of Qualitative Studies in Education* 26 (6), 629–633.

Littleton, K. and Howe, C. (eds) (2010) *Educational Dialogues: Understanding and Promoting Productive Interaction*. Abingdon: Routledge.

Littleton, K. and Mercer, N. (2013) *Interthinking – Putting Talk to Work*. Abingdon: Routledge

Mercer, N. (2000) *Words and Minds: How We Use Language to Think Together*. Abingdon: Routledge.

Phipps, A. (2013) Linguistic incompetence: giving an account of researching multilingually. *International Journal of Applied Linguistics* 23 (3), 329–341. (doi:10.1111/ijal.12042)

Phipps, A., Sitholé T. and Andrews, J. (2016) Words that nourish. Paper presented at the *Bristol Food Connections*, Bristol, 3 May 2016.

Phipps, A. (2015) Population healing: Languages, creativity and the extraordinary normality of migration. Paper presented at *Glasgow Centre for Population Health Seminar Series 12*, Glasgow, 21 October 2015. www.gcph.co.uk/assets/0000/5437/Alison_Phipps_Summary_final_final.pdf (accessed January 2020)

Tordzro, G. (2016) *The Calabash People*. https://vimeo.com/138059033 (accessed January 2020)

Wells, G. (1986) *The Meaning Makers*. London: Hodder and Stoughton

White, R., Fay, R., Kasujja, R. and Okalo, P. (2015) Global mental health: The importance of contextual sensitivity and appropriate methodologies. *MAGic 2015 Anthropology and Global Health: Interrogating Theory, Policy and Practice*. Brighton: University of Sussex.

Zwaan, K. (2010) Dutch court decisions and language analysis for the determination of origin. In K. Zwaan, P. Muysken and M. Verrips (eds) *Language and Origin: The Role of Language in European Asylum Procedures: A Linguistic and Legal Survey* (pp. 215–225). Nijmegen: Wolf Legal Publishers.

7 Translanguaging Beyond Bricolage: Meaning Making and Collaborative Ethnography in Community Arts

Jessica Bradley and Louise Atkinson

Introduction

In this chapter we reflect on a programme of collaborative arts-informed pedagogical activities, *LangScape Curators* (henceforth LS-C), to consider the concept of *bricolage* (Lévi-Strauss, 1962), its methodological implications and its potential affordances for developing understandings of translanguaging as dynamic multilingual and multimodal practice. LS-C was conceived as an outreach and research project to include young people's voices in research around multilingualism and cities (stemming from research conducted by the University of Leeds case study for the 'Translation and Translanguaging' project, (TLANG), Creese *et al.*, 2014–2018, see https://tlang.org.uk/). It was initially funded and supported by the University of Leeds' Educational Engagement team and therefore viewed as outreach with young people under the auspices of the university's widening participation activities. It has now been developed as part of a further collaborative research programme with young people, focusing on languages, creative inquiry and research into the linguistic landscape (Bradley, 2019). The collaboration described in this chapter is three-fold:

(1) It incorporates co-production across disciplines, with applied linguists working with artist-researchers.
(2) Researchers and artists work with professional colleagues from across the university to develop research which both informs and is informed by outreach and access programmes.

(3) Researchers, artists and professional colleagues work across sectors with educational professionals and young people.

In this chapter, using examples taken from activities across LS-C – which included making group collages, creating 'zines', writing stories, and building sculptures – we develop our understanding of translanguaging practices within the context of arts-based learning activities which have a broad focus on language and communication. The chapter follows on from a previous publication which focused on the project methodology (see Bradley *et al.*, 2018). We explore how bricolage might be considered as a starting point for a conceptual framework for transdisciplinary pedagogical activities, positioned at the intersection of research, practice and engagement and the role of collaboration in these. We also reflect on translanguaging's 'transformational' affordances in line with what might be considered a conceptual shift from a language-centred approach towards the *semiotic* and *embodied*.

Arts-Based Methods as De-Centring Language

The framing methodology for LS-C is broadly situated within arts-based learning, and we ask how creative practices might be woven into our 'translanguaging lens', therefore seeking to contribute moving towards the 'de-centring of language' (Thurlow, 2016: 503, see also Harvey *et al.*, 2019) while still providing a way for the 'expanded complex practices of speakers who could not avoid having had languages inscribed in their body' (García & Li, 2014: 18) to be made visible (as well as audible). In this way, we situate this study within a creative turn in applied linguistics (Bradley & Harvey, 2019), which moves away from the logocentric representationalism which privileges *language* as the mode in which we know the world (e.g. Thurlow, 2016). Our work is therefore grounded in the challenge of all 'text'-centred analysis, described by Crispin Thurlow in the context of 'queering' Critical Discourse Studies and as creating obstacles for understanding less visible identities and trajectories:

> And, in spite of a capital-C critical concern for the oppressed (people) and the hidden (ideologies), our textualism – our centring of texts and transcripts – leaves us struggling to read between the lines, to understand the gaps and the traces, the unspoken and the unspeakable. (Thurlow, 2016: 487)

Following Lynn Butler-Kisber (2010), we consider the hybrid research and outreach activity described within this chapter as 'arts-informed inquiry' (2010). As Butler-Kisber explains, 'arts-informed inquiry uses various forms of art to interpret and portray the focus of the particular study' (2010: 8). It is important to state that while LS-C uses creative methods to inform the research, it was not originally based on art (2010: 10); we therefore seek to highlight a nuance here. Butler-Kisber suggests that researchers using the arts in this way use 'arts-informed' to describe their practice, with artists working this way using 'arts-based'. As this project arises from research grounded in applied linguistics, we describe it as arts-informed. However, as the research develops further, the role of the artist becomes more central and future iterations of this inquiry foreground the artistic products – the objects created – themselves, therefore aligning more with the concept of 'arts-based' (see Atkinson & Bradley, 2019). The research we describe is collaborative, with artists and researchers working together at all stages to create the project. There is scope for critical debate as to the terminology used when working collaboratively and we have opted to use both terms in this chapter to show different perspectives.

Context: Young People and Inner-City Semiotic Landscapes

LS-C (2015-2017) was developed and delivered in the northern city of Leeds, UK, by the authors (Jessica and Louise) with colleagues Emilee Moore and James Simpson. Over the course of the project we worked with approximately 50 young people, aged between 11 and 14 years. The impetus for developing the work stemmed both from Jessica and Louise's experience of developing and delivering arts-informed and language-based educational engagement programmes within and outside the university and from the team's desire to continue to develop collaborative partnerships and externally-facing activities with young people around research into communication across the city. However, it is important to note that we did not consider LS-C solely as 'outreach' work or as public dissemination of research. These activities are sometimes positioned as a one-way transmission in which expert research findings are shared with the public, usually towards the end of the research project. Instead, we viewed it as research practice in itself, as 'co-production' (Bell & Pahl, 2018; Facer & Pahl, 2017; McKay & Bradley, 2016) between researchers, artists and young people. We sought to further explore how methodologies around linguistic and visual ethnography (Copland & Creese, 2015) could develop towards collaborative engagement with children and youth (Campbell & Lassiter, 2010; Hackett, 2017; Pahl, 2014), enabling us to pay attention to the relational and dialogic processes inherent within this kind of research activity (Siry, 2015: 151).

To create LS-C we worked with the university's educational engagement and outreach office and with an educational charity with whom the university is closely associated, whose strategy is to develop educational pathways for children and youth living in disadvantaged areas. The charity's two Leeds-based educational centres formed the project sites. One of the two centres is based in Harehills, within the city suburb in which the research team had been working for a number of years: Harehills was also the focus for two of the four phases of work undertaken for the Leeds case study of the TLANG project (see Baynham *et al.*, 2015, 2016; Simpson, 2011). In conjunction with the centre teams, we worked to develop a three-day programme of workshops and activities and then a further two-day summer school hosted by the university.

Ethnography, as an underpinning theoretical and methodological concept, was central to the workshop design. In developing the programme, we drew from our own individual research projects – linguistic ethnography and in fine art – both of which take ethnography as a central approach (cf. Grenfell & Pahl, 2018). Jessica's doctoral research (2018) focused on multilingual and multimodal communicative processes in street arts production and performance, and she asked to what extent *translanguaging* can be extended to incorporate the multimodal and performative practice of the street artists with whom she was working (see also Bradley & Moore, 2018). This research also informs the LS-C project in terms of its design, in itself, a bricolage, in Jacques Derrida's terms, 'borrowing from one's textual heritage whatever is needed to produce new and different texts, with an emphasis on intercultural borrowing for the purposes of textual construction' (Derrida, in Berry, 2015: 79). Louise's practice-based doctoral thesis (2016) took a theoretical framework of anthropology to consider aspects of appropriation in visual art. Her research focus was on how audiences might be engaged more effectively and collaboratively through anthropological approaches, while also addressing the significant problematic aspects related to cultural appropriation. The collaboration between Jessica and Louise built initially on these areas of alignment and through a growing interest in the interdisciplinary affordances of linguistic landscape research (e.g. Blommaert, 2013). Although it is not within the scope of this chapter to provide an extensive review of research in this area, it should be noted that we considered our work to be research and practice into visual and linguistic city landscapes, with the role of language understood broadly (Pennycook, 2017).

Within the programme of workshops, the young people were given an introduction to basic research skills, focusing on what research *is* and what linguistic landscape researchers might *do*. They were shown current examples of linguistic landscape research, including from the Leeds area undertaken for the TLANG project, and methods they might use to gather data. They were also given an overview of simple research ethics and the kinds of issues that might arise when collecting data in the street (for example when taking photographs in public spaces and requesting interviews with members of the public). The participants then went out into the 'field' in small groups, each with a member of centre staff, to explore their neighbourhoods, using photography, film and interviews as core data collection methods.

Using their initial findings, they then engaged with a diverse range of creative methods, led by practicing artists with interdisciplinary expertise. Later in the chapter we focus on three of these activities, using bricolage as an overarching theme: group collage, individual 'zines', and group collage sculptures. All the creative activities described in the chapter served to enable the participants to synthesise, analyse, re-present and communicate the data they collected in the field.

A Cautious Extension: Translanguaging, Multimodality and Artistic Practice

The introductory chapter to this book includes a discussion of translanguaging, a concept which is taken up in different ways by this volume's authors. Taking a translanguaging perspective on an arts-informed learning project of this kind is something we approach both cautiously and critically. Li Wei suggests that its application across a broad range of practices could imply that translanguaging is 'any' kind of communicative practice that is 'slightly non-conventional'. In arguing for the validity of translanguaging as a term in an era of 'post-multilingualism' he describes the context in which we find ourselves living and researching:

> multiple ownerships and more complex interweaving of lang-uages and language varieties, and where boundaries between languages, between languages and other communicative means, and the relationship between language and the nation-state are being constantly reassessed, broken, or adjusted by speakers on the ground. (Li, 2017: 14–15)

Defining 'ownerships' of language or any communicative mode is a complex and contested issue. In some cases translanguaging focuses on the individual and their own repertoire (see Otheguy *et al.*, 2015), and, in this sense, ownership rests with the individual. Others, for example Pennycook (2017), ask how considering linguistic landscapes *as repertoire* (e.g. Busch, 2012; Gumperz, 1964; Rymes, 2014), as Durk Gorter and Jasone Cenoz (2015) suggest, might inform understandings of translanguaging as spatial practice, extending beyond the individual. According to Gorter and Cenoz (2015), positioning the linguistic landscape as a 'multilingual and multimodal repertoire' (2015: 71) and applying a translanguaging lens to the study of these repertoires, has the potential to open up the field of multilingualism. Likewise, we posit that co-productive methodologies which centre on visual arts have significant affordances for linguists working in this area, as we shall discuss in this chapter.

Translanguaging, following Li, is therefore a useful lens for understanding 'what language is for ordinary men and women in their everyday social interactions' (2017: 15). Taking language from the individual's perspective offers a way to understand how 'ordinary men and women' (and, by extension, ordinary children and ordinary young people) view their own communicative practices. To what extent do they understand and claim ownership of their own repertoires, and to what extent are they are enabled to understand and claim any ownership? Likewise, how might everyday engagement with linguistic landscapes as repertoire affect this? This question reflects our decision to develop this research collaboratively with young people. We argue that in its expansion to incorporate the multimodal (see Blackledge & Creese, 2017; Bradley & Moore, 2018; Kusters *et al.*, 2017), *a translanguaging approach* enables an extension towards the non-linguistic. As Li puts it:

> Language, then, is a multisensory and multimodal semiotic system interconnected with other *identifiable* but *inseparable* cognitive systems. Translanguaging, for me, means transcending the traditional divides between linguistic and non-linguistic cognitive and semiotic systems. (2017: 15)

Multimodality is, according to this conceptualisation, integral to translanguaging. Bringing together social semiotics and translanguaging (see Sherris & Adami, 2018) can enable researchers to consider the multimodal affordances of translanguaging in understanding ordinary communication by ordinary people in ordinary contexts. There is still much progress to be made in establishing new analytical models which foreground translanguaging and this is the focus of current research in the creative arts sector by members of the LS-C project team. For example, Emilee Moore uses musical notation to demonstrate the resemiotisation of a poem and Jessica Bradley describes how puppets and objects created by street artists for outdoor performance become the objects of analysis rather than simply the context for the surrounding linguistic analysis (Bradley & Moore, 2018, see also Moore & Bradley, 2020).

In seeking to extend the translanguaging lens towards arts practice, encompassing multimodality, we are aware that there are gains in terms of broadening our perspectives and our developing theories. But we are equally aware that we risk losses (cf. Kress, 2005). Debates around the current multiple (and multiplying) theoretical conceptualisations of dynamic multilingualism abound (see Jaspers, 2018, for a discussion of translanguaging's 'transformational limits'). Alastair Pennycook refers to this as the 'trans-super-poly-metro' movement (2016: 201). As Li states, 'translanguaging seems to have captured everyone's attention' (2017: 9) and we, as authors, are conscious of the challenges of stretching and extending translanguaging beyond spoken and written language. The origins of translanguaging and its take-up in bilingual education (e.g. García, 2009; García & Li, 2014) are embedded in social justice and bilingual education rights, and any application of the concept should be mindful of its roots in addressing serious issues around language inequalities.

Arguably there is significant scope in situating a discussion of translanguaging within a creative arts context. The arts too, in education and more widely in social life, are a site of conflict. In the UK, shifts in educational policy and curriculum changes have affected arts subjects in the education sector. Introducing arts-informed learning as a methodology for our transdisciplinary educational workshops also therefore seeks to address some of the challenges faced by arts subjects in schools in the UK (as exemplified in cancelling and then reinstating Art History at Advanced Level, see Weale, 2016) and internationally and to demonstrate how arts-informed research can be used across diverse subjects. This extends the collaboration across disciplines and demonstrates clear areas of application at policy level for this kind of transdisciplinary research. As our research developed we wanted to capture empirical evidence of how arts-informed methods and working with artists might build spaces for young people's creative linguistic practice. How might young people from diverse backgrounds, living and studying in superdiverse wards of inner-city Leeds, be empowered to draw from across their full communicative repertoire within the context of our workshops?

From a methodological perspective we were interested in discovering how different ways of working collaboratively might enable new opportunities for translanguaging spaces (e.g. Li, 2011; Bradley & Simpson, 2019) to be created. These we understand following Li (2011):

> a Translanguaging Space, a space that is created by and for Translanguaging practices, and a space where language users break down the ideologically laden dichotomies between the macro and the micro, the societal and the individual, and the social and the psychological through interaction. (2011: 23)

We considered how our potential findings might shed light on new ways in which 'creative translanguaging spaces' could be brought into the classroom and foregrounded as pedagogical practice. But, translanguaging also underpinned the project design and the ways in which we as artists and researchers worked with the participants. We maintained an active focus on the idea of opening up translanguaging spaces and on engaging with the ideas the young people had about their everyday communicative repertoires. We observed multiple pedagogical practices by centre staff in encouraging the participants to draw on their communicative repertoires and have pride in their translanguaging practices. By taking translanguaging as epistemology, these workshops also developed new ways of considering collaboration and co-production in research, in practice and in engagement.

Bricolage

Bricolage is from the French verb 'bricoler' and translates roughly to English as 'to tinker' or 'to patch together'. Put simply, the researcher as bricoleur always 'borrows from other texts' (Berry, 2015: 103). Claude Lévi-Strauss's bricoleur 'works with his hands and uses devious means compared with those of a craftsman' (1962: 16–17). However, Lévi-Strauss is clear: although a bricoleur will use whatever is to hand to perform their task, the repertoire at their disposal is always limited. If we are to fashion together whatever we can from a limited repertoire, we can assume that what we create is also limited, by default. Focusing on writing in online contexts, Myrrh Domingo, Carey Jewitt and Gunther Kress suggest that bricolage in research might be considered by some as incoherent:

> We might feel that a 'bricolage', assembled casually on a beach from bits of flotsam and jetsam is incoherent. Yet its frame – some bits of branches and driftwood – around the collection of elements, can immediately suggest the potential to 'read' meaning into the ensemble. (Domingo *et al.*, 2015: 258)

And yet, as they state, research through bricolage is not automatically incoherent. The above authors differentiate between two broad kinds of bricolage in the context of online writing: a 'semiotic entity' for which the author has endeavoured to produce coherence for the reader (or observer) and one for which the audience is tasked with piecing the meanings together and finding their own coherence. Both can be considered as bricolage and both draw from a particular spatio-temporal repertoire. This relates directly to intentionality, and how the author of the work (be that a piece of art, a book, an online text) perceives the interaction between herself, the work and the audience.

Norman Denzin and Yvonna Lincoln (1994, 2000, 2005) describe the potential of bricolage as 'addressing a growing concern about what counts as and how to do research in an age of postmodernism, other postdiscourses and digital technologies' (in Berry, 2015: 81, see also Butler-Kisber, 2010). Translanguaging (alongside associated concepts, including metrolingualism [e.g. Pennycook & Otsuji, 2010] and polylanguaging [Jørgensen, 2008; Jørgensen *et al.*, 2011]) ask similar questions of us as researchers. As argued earlier in this chapter, when we take translanguaging as epistemology, we require new analytical tools and new approaches. Within this framework, it is insufficient for a researcher to go into the field, take photographs of the linguistic landscape around them, and then analyse according to their own interpretations. There is a need for engagement and collaboration: and these form part of the developing bricolage.

Collage as Practice and Theory

A critical approach to the notion of bricolage therefore underpins LS-C. The starting point for these is a collage activity, developing from Louise's own artistic practice. As part of her ongoing investigation, she created a series of collages, named Place Myths (2017, see www. louiseatkinson.co.uk/artwork/place-myths/). These were intended to evoke associations of space and place through using images taken from holiday catalogues interspersed with bold blocks of colour. Although the images were not based on real places and were not intended to represent any particular location, the use of colour and image aimed to conjure up feelings and memories of past tourist experiences. Collage in the context of LS-C explores the imagery and material culture of place – the linguistic landscape – through a variety of media, including drawing, collage, textiles and sculpture.

The term 'to collage' traditionally describes the act of grouping a selection of paper, 'found' photographs, and other ephemeral media and adhering onto a surface. Collage is entrenched in the art history canon through the work of early 20th century modernists, and its use has expanded within visual communication in general, from the political and social commentary of Hannah Hoch and John Heartfield, to the album cover designs of Peter Blake. In Louise's work, collage represents practice as research, and in developing the workshops based on her own creative practice, she works with arts-based methods (following Butler-Kisber's distinction outlined earlier).

The use of collage has also contributed to the development of visual inquiry as a participant research method, in part due to its perceived accessibility and a shift towards the creation of non-linear narratives in research. Butler-Kisber (2010) describes 'collage inquiry' (see also Prasad, 2018) as addressing dissatisfaction in qualitative research with 'traditional forms of representation' (2010: 102). She writes about collage *in inquiry, as reflection, as elicitation* and *in conceptualisation*. As she and Tilu Poldma argue, the use of collage making in research has multiple affordances:

> These new, arts-informed modes of inquiry mediate different kinds of understandings grounded in direct experiences, expand the possibilities of diverse realities, counter the hegemonic and linear thinking often associated with traditional research, increase voice and reflexivity in the research process, and create more embodied and accessible research results (Butler-Kisber & Poldma, 2010: 1)

But there is a risk in denying these methods a history or biography (Pool, 2018). In approaching LS-C from a professional practitioner background in which artist-led creative activity was embedded in learning, we envisaged that the creative methods used within the workshops should stem from existing professional arts practice. In this way, we aimed to reinforce the embodied and experiential nature of practice as research (and research as practice). Both authors also had experience of the affordances of creative practice in producing research results which are accessible to wider audiences. Collage, therefore, facilitated the production of three further activities within the workshop programme: group collage, 'zines', and collage sculpture. These creative arts activities are now discussed in more detail.

The Starting Point: Collage

Once the young people had conducted their fieldwork in the streets surrounding the educational centres, they came back to the base to work in their groups to produce collages. Collage was selected as the starting point for the creative activities for a number of reasons. First, it represented an activity that allowed the participants to immediately start the process of inquiry, reflection, elicitation and conceptualisation. Second, collage acted as a catalyst for them to process what they had found during their excursions. Third, it enabled them to approach the creative arts activity without concerns about their own artistic ability. All participants – no matter what level they considered their expertise – were able to select images, cut out images and start to gather them together on a large sheet of paper.

As explained earlier in the chapter, the collage activity was framed in reference to Louise's own artistic practice. Participants were invited to analyse images of her Place Myth series. In so doing, the exercise aimed to elicit responses about types of place, for example, mountains, cities and beaches, in addition to visual links to ideas around temperature through the use of warm or cold colour palettes. Following this activity, they worked in small groups to use these ideas when considering the process of conducting fieldwork and their findings.

Each group was given a sheet of A1 cartridge paper and a series of A4 colour images of the photographs that they had taken during their fieldwork. They also had access to newspapers and magazines from which they could cut out letters and words that might resonate with the conversations that they had had during their neighbourhood walks and any interviews they might have been able to conduct with people in the streets surrounding the centres. Drawing on the Place Myth collages, the young people were encouraged to produce images that represented their research process and findings, using the materials provided.

A translanguaging perspective gave us an alternative lens for the activity. The young people approached the activity in different ways. Some focused on image, some focused on text while others brought together a mixture of the two. Participants also started to introduce texture and additional imagery into their work through the addition of coloured tissue paper and by using acrylic paints. Figures 7.1–3 are taken from the activities carried out in the east of the city. Figure 7.1 shows how photographs of terraced houses taken during the fieldwork exercise are juxtaposed with quotations from interviews carried out by the young people and extracts from books used during the creative writing workshop. A series of flags are depicted in the top right hand corner and the word 'Langscape' features centrally. The clouds show excerpts of 'data' from the fieldwork exercise conducted by the young people. The terraced houses are typical of the area, but the green hills at the bottom of the collage are unexpected, although the educational

Figure 7.1 Collage example

Figures 7.2 and 7.3 Collage examples

centre was located by a park. Where images were used as a reference for drawn imagery, participants used their smartphones to find additional inspiration, therefore incorporating an element of digital literacy into the process.

In Figures 7.2 and 7.3 the participants have used paints to recreate real and imagined elements of the landscapes they investigated. Both figures repeat the green motifs seen in Figure 7.1, and each show a road. Figure 7.2 depicts the building in which the centre is based and a sign for the centre itself. In Figure 7.3 the road cuts across the image of a house, taken across the road from the centre. The words of some signs are cut up, creating a sense of known and unknown, familiar and unfamiliar.

The creation of the group collages enabled conversations about the research findings and facilitated discussion about prevalent themes within the work. These themes included aspects of community, language, ideas around nationality, and a critical analysis of the use of public space.

Extension and Communication: 'Zines'

When planning for the second set of workshops in the south of the city, we decided to create 'zines' (e.g. Lovato, 2008) to extend the young people's collaborative collages and as a way for participants to build on the themes identified through their fieldwork and in their data. Zines come from radical and creative arts practice in which artists and writers create their own 'low budget' publications to construct personal narratives. These booklets – often self-published (Lovato, 2008) – are handmade and in most cases produced in small editions. They utilise collage, drawing and handwritten content, in order to be easily photocopied and distributed. As Anna Poletti writes, zines and the zine writing, making and sharing community constitute 'a form of alternative media, a subculture of storytelling and knowledge sharing' (2005: 184). Zines therefore offered a productive format for our participants to catalogue and communicate their research findings. The zine, as a medium incorporating text and image in a sequential format, encouraged the young people to think about how audiences might 'read' their research in the form of 'fieldnotes' or even as a guide to the local area.

To produce the zine, participants were shown a simple folding technique to produce an 8-page booklet. They were then given the brief to produce an artwork in response to how they might represent the linguistic landscape of the surrounding area to a visitor, drawing from their own ethnographic research in the street. Participants drew on their own personal experiences of living and studying in the area, including from their own family histories and faiths, to share aspects of the food, activities and languages in their neighbourhood. Figures 7.4–7 below show the participants creating the zines and examples of the finished objects.

The young people focused on the site of the workshops as being an area which needed to be promoted. They drew on the diversity – linguistic and cultural – of the area, as explored during the fieldwork exercise. The globe featured as an image across a number of the zines (see Figure 7.7), connecting the local with the global, again linking to their explorations of the area and their lived experiences of it as multilingual residents of Leeds.

Incorporating the 3D: Collage Sculpture

Following the first stage workshops at the educational centres, we invited the groups with whom we had worked to the university for a two-day summer school. For this we were immersed in a different semiotic landscape: that of the university campus.

The summer school built on the creative arts workshops delivered previously and incorporated an aspect of sculpture. The participants at the summer school were, in the main, the young people with whom we had worked initially who had chosen to continue with the project. We wanted to build in a 3D aspect to the workshops and to encourage the participants to think about architectural considerations and how these might be taken into account in linguistic landscape research. As with the earlier programme, the young people went in groups to explore the surrounding linguistic landscapes, with the university campus as their site of critical analysis.

Figures 7.4, 7.5, 7.6, 7.7 Examples of zines

The basis for the collage sculpture activity was the printed photographs from these explorations. But in this instance, they also had to produce a 3D model on which to adhere their collage materials. The participants began by separating into small groups. Then, guided by Louise and supported by project staff, they worked to create a net of the 3D object, in a shape they chose as a group, based on the semiotic landscape of the campus. After they had constructed their models, they started to select and attach images around particular themes, including signage and architectural features. The models ranged from simple geometric shapes to complex renditions of university buildings incorporating staircases and, in some cases, stilts. The young people then presented their sculptures (see Figure 7.8) to the group and academics from the university, giving explanations for their choice of shape and for the selection of images and text.

Figure 7.8 Example of collage sculpture

The sculpture activity extended the collage work done previously to enable the young people to play with shape and structure, therefore responding to the architectural features of the linguistic landscape of the university. The shared making process required negotiation of which observations would be interwoven to create the sculptures.

Discussion

This chapter has explored the LS-C project as a process of collaboration between applied linguistics researchers and creative practitioners. In considering the different arts activities we have shed light on what these kinds of artistic artefacts and the processes involved in their creation tell us about translanguaging practices. The purpose of highlighting these methods is to demonstrate how an extension of our lenses as applied linguists through working collaboratively can further develop our understandings of communication as entangled with space. Bricolage, as a conceptual framework for this project, underpins the three activities shown here in multiple ways. First, as methodology, in bringing together Louise's creative practice with applied linguistics research. Second, as a process of engagement by young people with research and as collaborative research practice. Third, as a conceptual lens for considering how ordinary people draw on their repertoires to inquire about their own communities and localities, to reflect on their findings, to elicit new understandings through their inquiry and to conceptualise their understandings. The chapter focused on a series of artefacts, or objects, created over the course of the collaborative process.

However, this also moves us beyond bricolage. For every object a transformation took place. Each object was created collaboratively, with participants working with the materials they had to hand. The activities were designed in a way to encourage the young people to explore and to create as a process of transforming their own understandings of the linguistic landscapes around them. The transformation is therefore considered within new objects which represent the process of inquiry.

Participants used creative practice to synthesise, analyse and communicate their research findings from their communities, making visible the 'expanded complex practices of speakers who could not avoid having had languages inscribed in their body' (García & Li, 2014: 18), as previously cited, and the workshops aimed to provide a way to enable us, together with our participants, to overcome some of the barriers around reading 'between the lines', understanding 'the gaps and the traces' and 'the unspoken and the unspeakable' (Thurlow, 2016: 487).

In taking inner-city linguistic landscapes as our point of departure, the LS-C project, as arts-based and arts-informed, used multiple and diverse arts practices and methods to explore young people's understandings and knowledges of the communities in which they live and study. In this chapter we have considered the ways in which these activities – both as *creative* practice and as *research* practice – can be considered as iconic examples of multimodal translingual practice. We suggest that arts-based learning activities, as objects of arts-informed analysis, reveal new insights into how we deploy their communicative repertoire. The kinds of data we have described here are often seen as tangential to audio- or video- recorded data when conducting research into communication. We therefore argue that making the collaborative processes involved in research visible and disrupting traditional analytical processes is an important part of understanding dynamic communicative practice. LS-C also demonstrates the affordances of the linguistic landscape as a fruitful and creative catalyst for transformative collaborative research.

Using Lévi-Strauss' concept of bricolage, we have considered the artistic practice of collage and its application to the sociolinguistic concepts of repertoire and translanguaging. In setting out these examples as objects of analysis, we have sought to identify where and how artistic practice and dynamic multilingual practice might intersect through collaboration. By focusing on empirical multimodal data collected during the workshops, namely the art works produced by the young people, we have demonstrated how the concept of bricolage can be both harnessed and problematised as underpinning pedagogy. We have also set out how openness to bricolage in transdisciplinary arts-based practice, research and engagement, enables us to re-imagine and co-create linguistic realities. In doing so we challenge traditional ideas around bricolage, and suggest that the young investigators of the linguistic landscape are re-making and re-creating their multimodal worlds, using 'whatever is at hand'. However the young people's repertoires are not limited in the ways described by Lévi-Strauss. Instead, they are recreated and remade, in a process of constant renewal. The sets of creative and communicative tools deployed by the young people as they created the objects were open and contingent.

As with Lévi-Strauss's bricoleur, we, as chapter authors have drawn from a wide-ranging but finite repertoire. We write this chapter after the project has ended and we are conscious that we have analysed and theorised these artistic objects, created within the context of a collaborative ethnographic project, working together. We recognise that this chapter therefore models what Shirley R. Steinberg describes as a 'tentative interpretation' (2015: 111). We would like to thank the young LangScape Curators for allowing us to tentatively interpret their art works in this way.

Acknowledgements

The LS-C project was funded by the University of Leeds Educational Engagement Social Sciences Cluster. The authors would like to thank Steven Gleadall from the University of Leeds and the young people who participated in the project.

References

Atkinson, L. (2017) Place myths. www.louiseatkinson.co.uk/artwork/place-myths/, (accessed January 2020)

Atkinson, L. and Bradley, J. (2019) Art education as communicative repertoire: Questions for creativity in applied linguistics. Paper given at Annual Meeting of British Association for Applied Linguistics, August 2019

Baynham, M., Bradley, J., Callaghan, J., Hanusova, J. and Simpson, J. (2015) Translanguaging business: Unpredictability and precarity in superdiverse inner city Leeds. *Working Papers in Translanguaging and Translation* 4. https://tlang754703143. files.wordpress.com/2018/08/translanguaging-business.pdf

Bell, D.M. and Pahl, K. (2018) Co-production: Towards a utopian approach. *International Journal of Social Research Methodology*, 21 (1), 105–117.

Berry, K.S. (2015) Research as bricolage: Embracing relationality, multiplicity and complexity. In K. Tobin and S.R. Steinberg (eds) *Doing Educational Research: A Handbook* (2nd edn) (pp. 79–110). Rotterdam/Boston/Taipei: Sense Publishers.

Blackledge, A. and Creese, A. (2017) Translanguaging and the body. *International Journal of Multilingualism* 14 (3), 250–268.

Blommaert, J. (2013). *Ethnography, Superdiversity and Linguistic Landscapes: Chronicles of Complexity*. Bristol: Multilingual Matters.

Bradley, J. (2018) Translation and translanguaging in production and performance in community arts. (University of Leeds, unpublished thesis).

Bradley, J. (2019) *Multilingual Streets: Translating and Curating the Linguistic Landscape*. AHRC Open World Research Initiative.

Bradley, J. and Harvey, L. (2019) Creative inquiry in applied linguistics: Researching language and communication through visual and performing arts. In C. Wright, L. Harvey and J. Simpson (eds) *Voices and Practices in Applied Linguistics: Diversifying a Discipline* (pp. 91–107). Leeds/Sheffield/York: White Rose University Press.

Bradley, J. and Moore, E. (2018) Resemiotisation and creative production: Extending the translanguaging lens. In A. Sherris and E. Adami (eds) *Making Signs, Translanguaging Ethnographies: Exploring Urban, Rural and Educational Spaces* (pp. 81–101). Bristol: Multilingual Matters.

Bradley, J. and Simpson, J. (2019) Translanguaging in the contact zone: Mobility and immobility in inner-city Leeds. In K. Horner and J. Dailey-O'Cain (eds) *Multilingualism, (Im)mobilities and Spaces* (pp. 145–164). Bristol: Multilingual Matters.

Bradley, J., Moore, E., Simpson, J. and Atkinson, L. (2018) Translanguaging space and creativity: Theorising collaborative arts-based learning. *Language and Intercultural Communication, Special Edition, Bridging across languages and cultures in everyday lives: new roles for changing scenarios* 18 (1), 54–73.

Busch, B. (2012) The linguistic repertoire revisited. *Applied Linguistics* 33 (5), 503–523.

Butler-Kisber, L. (2010) *Qualitative Inquiry: Thematic, Narrative and Arts-Informed Perspectives*. London: SAGE.

Butler-Kisber, L. and Poldma, T. (2010) The power of visual approaches in qualitative inquiry: The use of collage making and concept mapping in experiential research. *Journal of Research Practice* 6 (2), 1–17.

Callaghan, J. (2015) *Changing Landscapes: Gipton and Harehills – A Superdiverse City Ward. Working Papers in Translanguaging and Translation* 7. https://tlang 754703143.files.wordpress.com/2018/08/changing-landscapes.pdf

Campbell, E. and Lassiter, L. (2010) From collaborative ethnography to collaborative pedagogy: Reflections on the other side of Middletown project and community-university relationships. *Anthropology and Education Quarterly* 41 (4), 370–385.

Copland, F. and Creese, A. (2015) *Linguistic Ethnography: Collecting, Analysing and Presenting Data*. London: SAGE.

Denzin, N.K. and Lincoln, Y.S. (eds) (1994, 2000, 2005) *The Handbook of Qualitative Research* (1st, 2nd, 3rd edns). Thousand Oaks, CA: SAGE.

Domingo, M., Jewitt, C. and Kress, G. (2015) Multimodal social semiotics: Writing in online contexts, In K. Pahl and J. Rowsell (eds) *The Routledge Handbook of Literacy Studies* (pp. 251–266). Abingdon: Routledge.

Facer, K. and Pahl, K. (2017) *Valuing Interdisciplinary Collaborative Research: Beyond impact*. Bristol: Policy Press.

García, O. (2009) *Bilingual Education in the 21st Century: A Global Perspective*. West Sussex: Wiley Blackwell.

García, O. and Li, W. (2014) *Translanguaging: Language, Bilingualism and Education*. Basingstoke: Palgrave Macmillan.

Gorter, D. and Cenoz, J. (2015) Translanguaging and linguistic landscapes. *Linguistic Landscape* 1 (1/2), 54–74.

Grenfell, M. and Pahl, K. (2018) *Bourdieu, Language-based Ethnographies and Reflexivity: Putting Theory into Practice.* London: Routledge.

Gumperz, J.J. (1964) Linguistic and social interaction in two communities. *American Anthropologist* 66 (6/2), 137–53.

Hackett, A. (2017) Parents as researchers: Collaborative ethnography with parents. *Qualitative Research* 17 (5), 481-497.

Harvey, L., McCormick, B. and Vanden, K. (2019) Becoming at the boundaries of language: Dramatic enquiry for intercultural learning in UK higher education. *Language and Intercultural Communication* 1–20. DOI: 10.1080/14708477.2019.1586912

Jaspers. J. (2018) The transformative limits of translanguaging. *Language and Communication* 58, 1–10.

Jørgensen, J.N. (2008) Polylingual languaging around and among children and adolescents. *International Journal of Multilingualism* 5 (3), 161–176.

Jørgensen, J. N., Karrebæk, M. S., Madsen, L. M. and Møller, J. S. (2011) Polylanguaging in superdiversity. *Diversities* 13 (2) www.unesco.org/shs/diversities/vol13/issue2/art2 (accessed January 2020)

Kincheloe, J. (2001) Describing the bricolage: Conceptualising a new rigor in qualitative research. *Qualitative Inquiry* (7) 6, 679-692.

Kress, G. (2005) Gains and losses: New forms of texts, knowledge, and learning. *Computers and Composition* 22, 5–22.

Kusters, A., Spotti, R., Swanwick, R. and Tapio, E. (2017) Beyond languages, beyond modalities: Transforming the study of semiotic repertoires. *International Journal of Multilingualism* 14 (3), 219-232.

Lévi-Strauss, C. (1962) *The Savage Mind.* Chicago IL: University of Chicago Press.

Li, W. (2011) Moment analysis and translanguaging space: Discursive construction of identities by multilingual Chinese youth in Britain. *Journal of Pragmatics* 43 (5), 1222–1235.

Li, W. (2017) Translanguaging as a practical theory of language. *Applied Linguistics* 39 (1), 9–30.

McKay, S. and Bradley, J. (2016) How does arts practice engage with narratives of migration from refugees? Lessons from 'utopia'. *Journal of Arts and Communities, Special Edition Arts, Activism and Human Rights* 8 (1–2), 31–46.

Moore, E. and Bradley, J. (2020) Resemiotisation from page to stage: The trajectory of a musilingual youth's poem. *Journal of Bilingual Education and Bilingualism* 23 (1), 49–64.

Otheguy, R., García, O. and Reid, W. (2015) Clarifying translanguaging and deconstructing named languages: A perspective from linguistics. *Applied Linguistics Review* 6 (3), 281–307.

Pahl, K. (2014) *Materialising Literacies in Communities: The Uses of Literacies Revisited.* London: Bloomsbury.

Pennycook, A. (2016) Mobile times, mobile terms: The trans-super-poly-metro movement. In N. Coupland (ed.) *Sociolinguistics Theoretical Debates* (pp. 201–216). Cambridge: Cambridge University Press.

Pennycook, A. (2017) Translanguaging and semiotic assemblages. *International Journal of Multilingualism* 14 (3), 269–282.

Pennycook, A. and Otsuji, E. (2010) Metrolingualism: Fixity, fluidity and language in flux. *International Journal of Multilingualism* 7 (3), 240–254.

Poletti, A. (2005) Self-publishing in the global and local: Situating life writing in zines. *Biography* 28 (1), 183–192.

Pool, S. (2018) Everything and nothing is up for grabs: Using artistic methods within participatory research. In K. Facer and K. Dunleavy *Connected Communities Foundation Series*. Bristol: University of Bristol/AHRC Connected Communities Programme.

Prasad, G. (2018) 'But do monolingual people really exist?' Analysing elementary students' contrasting representations of plurilingualism through sequential reflexive drawing. *Language and Intercultural Communication* 18 (3), 315–334.

Rymes, B. (2014) *Communicating Beyond Language*. New York: Routledge.

Sherris, A. and Adami, E. (eds.) (2018) *Making Signs, Translanguaging Ethnographies: Exploring Urban, Rural and Educational Spaces*. Bristol: Multilingual Matters.

Simpson, J. (2011) *Harehills ESOL Needs Neighbourhood Audit*. Leeds City Council, YoHRSpace.

Siry, C. (2015) Researching with children: Dialogic approaches to participatory research. In K. Tobin and S.R. Steinberg (eds) *Doing Educational Research: A Handbook* (2nd edn) (pp. 151–166). Rotterdam/Boston/Taipei: Sense Publishers.

Steinberg, S. (2015) Proposing a multiplicity of meanings: Research bricolage and cultural pedagogy. In K. Tobin and S.R. Steinberg (eds) *Doing Educational Research: A Handbook* (2nd edn) (pp. 111–132). Rotterdam/Boston/Taipei: Sense Publishers.

Thurlow, C. (2016) Queering critical discourse studies or/and performing 'post-class' ideologies. *Critical Discourse Studies* 13 (5), 485–514.

Weale, S. 2016. Art History A-Level saved after high-profile campaign. The Guardian, www.theguardian.com/education/2016/dec/01/art-history-a-level-saved-from-being-axed-after-high-profile-campaign. (accessed January 2020)

Zhu, H., Li, W. and Lyons, A. (2017) Polish shop(ping) as translanguaging space. *Journal of Social Semiotics* 27 (4), 411–433.

8 Telling the Stories of Youth: Co-Producing Knowledge across Social Worlds

Emilee Moore and Ginalda Tavares Manuel

Introduction

This chapter reflects on a process of collaboratively producing knowledge between a university-based teacher-educator | researcher (Emilee Moore) and a poet | researcher (Ginalda Tavares Manuel)[1] who came together through a Youth Spoken Word (YSW) poetry organisation called Leeds Young Authors (LYA) based in Leeds, UK. This process was embedded in a broader linguistic ethnographic project led by Moore, with the organisation as its focal site, which formally took place over a period of 20 months from December 2015 to July 2017. The ethnographic project emerged from an interest in understanding the socially and educationally transformative potential of YSW, as both a powerful artistic and pedagogical practice, and as a transnational youth culture connecting diverse young people across the globe. YSW organisations aim to empower youth to use their ideas, their words, their voices, their bodies and their emotions as catalysts for personal development, critical learning and social change (e.g. Ibrahiim, 2016; Yanofsky *et al.*, 1999). As a teacher-educator | researcher, Moore's goal was to learn from experiences and expertise developed outside of mainstream education in contributing to socially transformative pedagogical practices involving language in schools. Tavares Manuel was a teenage member of Leeds Young Authors at the time of the research.

One of the key conceptual notions framing the ethnographic research was that of translanguaging, as particular attention was paid to fluid practices that spanned oral, visual, embodied, and spatial modalities (Blackledge & Creese, 2017; Bradley & Moore, 2018; Li, 2017; García & Li, 2014). Poetry is a particularly interesting practice for studying such semiotic processes. As van Leeuwen (1999: 5) writes about poetry: '[...] things work differently. No hard and fast rules exist. Any bit of language you might lay your hands on could come in handy for the semiotic job at hand, whether it is grammatical or not, whether it represents a standard variety of English or not.' Following García and Li (2014), such fluidities provide new ways of thinking about language in contexts of diversity and are opportunities for transforming subjectivities, as well as social and educational structures. Through fine-grained analyses of multimodal data (e.g. Goodwin, 2000; Mondada, 2016; Norris, 2004), the research aimed initially to advance the theoretical and methodological bases of translanguaging studies and to offer a deeper understanding of how resources available in language, bodies, objects and spaces combine to construct meaning (see Bradley & Moore, 2018; Moore & Bradley, 2020).

However, as often happens in ethnographic research, the ethnographer's role quickly moved beyond that of simply 'researcher', to that of 'member' of the organisation, as the needs of the youth and the organisation also changed. As part of this process, the ethnographer's understanding of what counted as knowledge production in the ethnographic research was also modified over time. Following authors such as Pahl (2014: 48), 'the way in which the collaborative space of inquiry that crosses the boundaries of arts practice, ethnography and education can open up new epistemological spaces, that in turn, listen to meaning makers' emerged as a possibility for enquiry. Taking the research in this direction did not deflect the focus from translanguaging, but rather translanguaging provided a robust and flexible lens that afforded new ways of seeing the dialogical production of knowledge and for engaging participants in the research in mutually beneficial ways. As a first step towards such co-production (McKay & Bradley, 2016; Pahl, 2014), in this chapter we describe a process of co-reading, co-interpreting and co-writing ethnography (Lassiter, 2005) we shared. We shall be moving between the first and third person in referring to ourselves, in order to take joint ownership for the text.

'Bleach'

Following the above introduction to the research that brought us together, we now introduce Gina's poem, 'Bleach'. We then present an overview of how our conversations began.

Bleach
A poem by Ginalda Tavares Manuel

This is the truth
I found lightening cream once
While scavenging for sweets
Or loose change at the bottom of my mother's bag

White cream
In a small white package

Even with a developing mind, I did not have to read the
Alpha hydraulic
Acidic ingredients
To know that
This was bleach

Prior to this
I had a problem with my melanin
Dark face
Light hands
Blackened eyelids
Due to scratching and skin rashes

'You're blik man'
'You're so black'
'Why you so dark?'
'You look burned gollywog'

I had a problem with my entire face
And used to cover it with long fringes
I'd long to be lighter like my Dad
And pray I never acquired my Mum's complexion

Through fear of feeling inferior
Beside, a lighter shade of black

I wanted to be light like Beyoncé
Like Tyra Banks
Because these days light is right
Too much melanin is displeasing

I still wanted to be Black
But not the shade of black that made me blend in
With the darkness

Later on in life
I learned
I was to love
Dark skin

And I grew
And noticed

Not a single blemish
Or spot
On my mother's face

Her skin is dark
Rich
Exotic
A rare pigmentation
She is a live creation

Skin that held
The freshest
Of deep rooted soil
Mixed with scorching sun rays
And blackberry juice

She
A very own
Walking
Africa

I love black skin
I love my skin

Skin that is blemished
Skin that has various shades and tones
Skin that has been whipped, torn and scorned
Skin that holds herbs, spices and wise tales
Black skin
Black skin that is beautiful

We, Gina and Emilee, first met on 12 January 2016, when Emilee started attending LYA's weekly writing sessions. Presenting herself as a university-based teacher-educator and researcher, Emilee told the attendees she was interested in how young people communicated with and beyond spoken language. She observed the sessions, took notes, and spoke up at times, conducting classic ethnographic participant-observation (e.g. Shah, 2017).

As weeks and months went by, it became clear that a lot of the poems written by Gina and the other poets in the group were about what it is like to be a teenager in Leeds and about social and political issues of concern to them. In particular, the youth often wrote about racism, as Gina did in her poem, 'Bleach'. This was the poem that Gina auditioned with, on 23 February 2016, for LYA's poetry slam team, and that Emilee later told Gina she was interested in analysing.

This text is part of a joint analysis. Gina's reasons for writing the poem and Emilee's reasons for wanting to use it in her research will become clearer as this chapter unfolds.

Just as Emilee started to consider how she could analyse Gina's poem with the knowledge and toolkit she had accumulated as a university-based researcher from fields such as Ethnography, Interactional Sociolinguistics, Literacy Studies or Multimodality, a colleague (Jessica Bradley) circulated an article published in the *New York Times Magazine* (Lewis-Kraus, 2016). The article discussed the controversies over Alice Goffman's book, *On the Run*, in which Goffman presents an ethnographic account of the lives of young Black men in Philadelphia, USA. Goffman was strongly criticised for failing to account for her positionality as a White academic in writing about people whose lived experiences as Black youth were quite different from her own.

The article dealt with an issue that has long been discussed in ethnography and in other research traditions, being that of representation (e.g. Van Maanen, 1995). The more Emilee learned from Gina and the other young poets through their writing and conversations, the less capable she felt – as someone who was not from Leeds (Emilee is originally from Australia, has lived for many years in Catalonia, and at the time was a visiting researcher in Leeds), was no longer a teenager, and had not been the target of racism – to adequately represent the meanings the young poets were trying to transmit through their creative practice.

'Storytelling Rights' and Collaborative Ethnography

At about the same time, Emilee was reading Amy Shuman's (1986) book *Storytelling Rights: The Uses of Oral and Written Texts by Urban Adolescents*. In the book, Shuman presents her research in an urban US high school, focusing especially on fight stories told by girls. One particular fight, involving a stabbing, was covered in the local newspaper, and the way the fight was represented caused the girls who were involved and witnessed it enormous anger. Basically, Schuman (1986: 123) concludes from the incident that '[outsiders had] no right (no entitlement) [...] unless they told the whole story'. Furthermore, she claims that 'The question of what constitutes the "whole story" involves the relationship between recontextualisation and entitlement' (1986). For Shuman, an adequately recontextualised story 'provides whatever contextual details are necessary' (1986), while entitlement is 'based upon a continual process of building shared understanding' (1986: 152).

Ethnographic approaches foreground subjectivities – or the contextual details that we and our participants see as relevant for reaching understandings from our particular positionalities, and *inter*subjectivities – or shared understandings between researcher and participants. Collaborative forms of ethnography also afford shared processes of material practice, including co-reading, co-interpretation and co-writing of ethnographic texts (Lassiter, 2005), and co-production (McKay & Bradley, 2016; Pahl, 2014), including with artists. Collaborative ethnography frames how Emilee decided to approach the analysis of Gina's poem, and this chapter is the outcome of an initial experiment with how that might work in practice, and what the potential of that process might be for us both.

Co-reading, Co-interpreting, Co-writing

When we decided to write this chapter together, we met up at the university coffee shop, and we recorded our first meeting. We did not know what our collaboration would involve exactly, just that we wanted to do something together. Our starting point was Gina's poem, 'Bleach'. The following (Fragment 8.1) is an extract from that first meeting, when we are discussing what brings us together, the prior experiences and expectations we have of doing qualitative research, how to work together collaboratively to tell the 'whole story' about Gina's poem, and how to each benefit from that process. Gina is represented in the transcript as GI. Emilee is represented as EM. The transcript employs a reduced set of Jefferson's (2004) conventions.[2]

Fragment 8.1
```
1.   EM: I'm doing re- so I'm doing research about you guys
2.       right?
3.   GI: hm. (.) and how we-
4.   EM: and so: the normal way of (.) doing things in
5.       academia right? (.) I think we spoke about this
6.       before (.) normally what university researchers do
7.       (.) is they go and they collect the data
8.   GI: yeah and [then they like]
9.   EM:          [and may:be]
10.  (.)
11.  GI: just process it without actually (.) sitting down
12.      with the
13.  EM: yeah
14.  GI: researchee?
15.  (.)
16.  EM: exact- yeah. [so it's like]
17.  GI:             [finding out] like kind of (.) first
18.      hand data really.
19.  EM: yeah. (.) so it's like (.) it would be my
20.      interpretation [of]
21.  GI:               [of] you. whereas [this is our-]
```

22. EM: [a story] you've
23.   produced really [yeah?]
24. GI: [yeah.] (.) oh yeah.
25. (.)
26. EM: and so the idea is to do that differently. (.) °you
27.   know?°
28. GI: [so that the person can actually tell]
29. EM: [we can tell the story together]
30. GI: yeah.
31. (..)
32. EM: and then also in telling the story what do you get
33.   from that? (.) you know what I mean?
34. GI: yeah.
35. EM: because I get- this is my job so I get something from
36.   it ((laughing) [no matter] what you know?)
37. GI: [yeah.]
38. GI: but then (.) as a young person (.) you get your
39.   voice heard.
40. EM: yeah. (.) what would you want from this? (.) I mean
41.   what would you want from resea:rch? (.) [what would
42.   you-]
43. GI: [just for it]
44.   to be looked at in a different way because (.)
45.   actually doing sociology in ehm college it was like
46.   was learning about the research and primary data and
47.   secondary data and (.) and all kinds of things that
48.   like so the researcher would just go and watch
49.   someone and they'd be like well the person's acting
50.   like this because I think they're acting like that
51.   for a reason [but it's like]
52. EM: [yea:h]
53. GI: you haven't really gone and asked the person. (.) so
54.   I think it's just about like recognising what the
55.   person is actually meaning so like you know [when
56.   you-]
57. EM: [yeah]
58. GI: you know when you're doing poetry
59. EM: yeah
60. GI: and you write something and then someone's
61.   interpreting it in their own way which [is okay]
62. EM: [yeah]
63. GI: but then they don't think about how what you actually
64.   [meant]
65. EM: [what it] meant
66. GI: what the meaning is behind
67. EM: yeah
68. GI: your poem or your line or [a word]
69. EM: [yeah]
70. GI: [so]

71. EM: [so] that's what I thought we'd do together
72. GI: yeah.
73. EM: so we'd interpret your poetry (.) but we'd also
74.   negotiate what we want from it (.) and we'll try and
75.   (.) [get-]
76. GI: [work] it.
77. EM: yeah?
78. GI: yeah. that sounds good.
79. EM: ((laughing)) yeah.

What is interesting in this fragment is how we dialogically construct the parameters for our shared research practice, and we begin to distinguish our practice from that of 'those other researchers' and 'those other researchees'. The way we build those categories together, and in particular the way the categories change over the two minutes that the fragment lasts, is summarised in Figure 8.1. Emilee modifies her research goal from 'doing research about you guys' (line 1), to 'do[ing it] together, so we'd interpret your poetry, but we'd also negotiate what we want from it' (lines 71–74). The aspiration is to alter the practices of other researchers, who position researchees as passive subjects according to the categorisation constructed in the dialogue, and other researchees, whose voices remain unheard. In sum, in the conversation we agree to tell Gina's story together, in a way that is beneficial to both of us.

The Learning Potentials of Boundaries

We are also interested in how we are both learning in this process of co-reading, co-interpreting and co-writing, and what sorts of dialogical settings support our collaborative knowledge building. As we both come into our shared research practice from different social worlds, the notions of boundary crossing and boundary objects are useful ones. These notions emerged in the cognitive and learning sciences and have more recently been incorporated into studies of language and discourse, coinciding with an increasing interest in objects and materialism, and how they impact talk and meaning making (e.g. Budach *et al.*, 2015, following Latour's Actor Network Theory, e.g. Latour, 2005). Following Akkerman and Bakker (2011), boundary crossing is the activity of going into a territory in which we are less familiar and feel less qualified, combining our repertoires of skills and tools from different boundaries of practice to achieve hybrid situations and understandings. Boundary objects are objects that '[...] both inhabit several intersecting worlds and satisfy the informational requirements of each of them' (Star & Griesemer, 1989: 393), thereby facilitating boundary crossing. Budach *et al.* (2015: 393) describe

Those other researchers	Our shared research practice	Those other researchees
	'I'm doing research about you guys' (line 1)	
'they go and they collect the data just process it without actually sitting down with the researchee' (lines 7–14)		
'it would be my interpretation' (lines 19–20)	'first-hand data' (lines 17–18)	
	'our story' (line 21) 'a story you've produced really' (lines 22–23) 'the idea is to do that differently' (line 26)	
	'we can tell the story together' (line 29)	
	'in telling the story what do you get from that?' (lines 32–33)	
	'this is my job so I get something from it no matter what' (lines 35–36) 'what would you want from research?' (line 41) 'for it to be looked at in a different way' (lines 43–44)	
'the researcher would just go and watch someone and they'd be like well the person's acting like this because I think they're acting like that for a reason' (lines 48–51)		
	'It's just about like recognising what the person is actually meaning' (lines 54–55)	
		'you know when you're doing poetry and you write something and then someone's interpreting it in their own way which is okay but then they don't think about how what you actually meant, what the meaning is behind your poem or your line or a word' (lines 58–68)
	'that's what I thought we'd do together, so we'd interpret your poetry, but we'd also negotiate what we want from it' (lines 71–74)	

Figure 8.1 Constructing our shared research practice

boundary objects as those which '(a) are able to move physically across contexts; and (b) are endowed with the ability to carry meaning'. They go on to claim that:

> Studying boundary objects, we argue, can enable the researcher to understand social processes in a more nuanced way. This is the case when the object-focus helps to identify meanings that occur simultaneously or are intersecting with each other, in one space or across spaces. Objects, then, become the reference point around which social meaning crystallizes, either in relation to a particular feature of object-design or as a form of human-object relation. Human-object relations, then, can prove a very fruitful lens to understand about broader social relations and issues of power. (Budach *et al.*, 2015: 393)

Inspired by these theoretical contributions, we present a second fragment from our first meeting, in which the way we define and begin to cross boundaries, and the significance of Gina's poem for this crossing, is clear. The fragment begins with Emilee bringing up some of her concerns about her difficulties in interpreting Gina's poem, 'Bleach', taking into account the subject position she constructs for herself in the interaction as a foreign, White, university-based researcher.

Fragment 8.2

1.	EM:	we're from different places (.) obviously: you know
2.		I'm- you're an insider in th- the poetry community
3.		and I'm not
4.	GI:	hm
5.	EM:	but I'm an insider in the research [community] you
6.		know?
7.	GI:	[oh yeah.]
8.	EM:	so maybe we're speaking:
9.	GI:	in different languages
10.	EM:	in complimentary ways right? (.) because we're doing
11.		both of those things at the same time? (.) ahm: (.)
12.		you know I'm not from he:re (.) I don't have an
13.		African background ((laughs))
14.	GI:	yeah
15.	EM:	I don't have a Portuguese background yeah? (.) but I
16.		have a:
17.	GI:	hm
18.	EM:	my background is in Australia
19.	GI:	yeah
20.	(.)	
21.	EM:	living in Spa:in (.) so I've got all of that.
22.	(.)	
23.	EM:	yeah (.) so (.) but I don't- do you think that

```
24.          that's anything that separates us? you know when we
25.          talk about [talking across cultures]
26.  GI:               [no because I think]
27.  EM:  I don't think so.
28.  GI:  no because I think it doesn't separa- separate us I
29.          think it might bring us like closer as people because
30.          in poetry you always have to research something.
31.  EM:  hm
32.  GI:  like (.) you can be asked to do a poem abou:t (.)
33.          post-colonial war [or something] like that
34.  EM:                       [hm:]
35.  EM:  ((laughs))
36.  GI:  and you have to go away and research [it]
37.  EM:                                        [yep.]
38.  GI:  because we don't know anything about it.
39.  EM:  yep.
40.  GI:  cause it's sometimes it's like writing from
41.          imagination (.) [and then] there's writing based on
42.          facts
43.  EM:                  [hm:]
44.  EM:  hm
45.  GI:  and that's what research does.
46.  EM:  hm
47.  GI:  you know what I mean and (.) also you research-
48.  EM:  that's true so you are a [researcher (.) that's true
49.          so true yeah.]
50.  GI:                           [yeah (.) and then (.) s-
51.          and in terms of re]searchers researchers can be poets
52.          too because
53.  EM:  yeah.
54.  GI:  you don't always have to present your- (.) ehm: I
55.          think poetry's all about being creative and you don't
56.          always have to present your findings on a piece of
57.          paper
58.  EM:  hm:
59.  GI:  you know what I mean. (.) and- what- the way we're
60.          doing it now we're doing it quite vo[cal quite]
61.          visual
62.  EM:                                          [hm hm]
63.  GI:  so that's poetry in itself it's quite creative and
64.  EM:  hm:
```

In the fragment, it is evident how Emilee, with Gina's dialogical support, initially sets up a lot of boundaries that she and Gina might be talking across in developing their shared research practice. She argues that she and Gina are 'from different places' (line 1), that Gina is an 'insider in the poetry community' (line 2), while Emilee is an 'insider in the research community' (line 5), that they are 'speaking in different languages' (lines 9–10), that Gina has an 'African background' (line 13), and a 'Portuguese background' (line 15), while Emilee has a 'background in Australia' (line 18), and has been 'living in Spain' (line 21).

In the second part of the fragment (from line 26), however, Gina very skilfully leads a dialogical reconfiguration of these boundaries. She argues that research and poetry, two practices that Emilee had constructed as being separately bounded to different social worlds, are in fact shared practices that cross over the boundaries set up in the interaction. According to Gina, poets are researchers and researchers can be poets. Therefore, the learning potential of crossing boundaries in this collaboration is not so much about Gina becoming a researcher, or about Emilee becoming a poet, but for Gina to learn to be a more university-type researcher, and for Emilee to become a more poetic one. The intersection of poetry and research is the point at which our shared research practice might emerge. Moreover, in terms of boundary objects, following Budach *et al.* (2015), Gina's poem is the material reference point around which shared social meaning might crystallise.

Developing Shared Social Meaning

In drawing this paper towards a close, we would like to come back to the analysis of Gina's poem, 'Bleach', which was what initially prompted our co-reading, co-interpreting and co-writing. In our first meeting, it was not Gina's edited written version of the poem that we worked with (i.e. the version that Gina chose to publish in this volume), but rather a rough and partial transcription of a recording of the poem as it was performed by Gina on 23 February 2016, when she auditioned for the LYA poetry slam team. We also had the video recording itself. These two material artefacts – the recording and the partial transcription – constituted our main boundary objects. The analysis we present will therefore not be of the entire poem as it appears in the written version presented at the beginning of this chapter, but rather we will focus on the first four stanzas, as they were the focus of our meeting.

Fragment 8.3

1. GI: this is the truth.
2. (..)
3. GI: I found lightening cream once:, while scavenging for
4.   sweets:
5. (.)
6. GI: or loose change at the bottom of my mother's bag.
7. (.)
8. GI: white cream in a sma:ll white package.

9.   (..)
10.   GI: even with a developing mind I didn't have to read that
11.   alpha hydraulic acidic ingredients to know that this was
12.   bleach.
13.   (.)
14.   AI: [(((clicking fingers))]
15.   GI: [prior to this]
16.   (.)
17.   GI: I'd a problem with my: melanin.
18.   (.)
19.   GI: dark fa:ce light ha:nds blackend eyelids.
20.   (.)
21.   GI: due to scra- scratching and skin rashes.
22.   (.)
23.   GI: ((stylised) you're blik man.
24.   (.)
25.   GI: you're so black.
26.   (.)
27.   GI: why you so dark?
28.   (.)
29.   GI: you look burned gollywog.)

One of the reasons that Emilee chose this particular poem to bring to the discussion with Gina was the impact that the words had had on her during Gina's audition. Another reason was Emilee's interest in translanguaging that we mentioned in the opening of this chapter. The latter curiosity meant that several aspects of the recorded poem stood out to her. For example, she noticed the non-verbal communication between Gina and her peer, Aida, in lines 14–15 – Aida shows her appreciation of Gina's words by clicking, as Emilee's ethnographic work with the YSW organisation revealed was a typical practice. Lines 23–29 stood out to Emilee especially – she was fascinated with how Gina brings in multivocality, or the coming together of multiple voices (Bakhtin, 1984), to express a meaning that she might not achieve otherwise. Gina stylises her voice, using a different speech variety than in the rest of the poem to mark reported speech ('you're blik man', line 23). Such use of different speech varieties was common among the young poets in their performances and formed part of their translanguaging repertoires. In Gina's poem, it also seems to index something that had been said to her by an unidentified other. Returning however to the problem of representation and of telling the whole story of the poem, this is where Emilee's analysis was insufficient without Gina's co-interpretation.

Indeed, Emilee initially understood the second voice that Gina brought into her performance of 'Bleach' to be that of someone in particular – a certain person that had spoken to Gina at a specific time, in a specific place. Emilee was interested in knowing more about that event, and the others that Gina's poem makes reference to (e.g. finding skin lightening cream in her mother's bag), and a lot of Emilee's questions in our meetings probed into this aspect. The following extract from a longer stretch of conversation during our first meeting exemplify this. The fragment begins with Emilee asking Gina where her poem comes from, and what she is trying to say with it.

Fragment 8.4

1. EM: where does it come from?
2. (.)
3. GI: it's real.
4. EM: what are you trying to say with it?
5. (.)
6. GI: that we are all beautiful and we shouldn't have to change
7.   our skin colour to
8. EM: hm hm
9. GI: fit in with someone else's perception of what's beautiful
10.   and what's not.
11. EM: hm.
12. GI: °you know what I mean? it's like°
13. (.)
14. GI: it's like you always follow your parents' examples and then
15.   you find your mother with that kind of cream you're like
16.   (.) so (.) what does [this mean? (.) if she's trying-]
17. EM: [when did it happen?]
18. GI: yeah (.) if she's trying to change that (.) her skin (.)
19.   then why- why- (.) am I trying to change mine?
20. EM: hm
21. GI: like I mean like I'm uncomfortable in my skin so
22. EM: hm
23. GI: you know (.) yeah (.) that's what it's just trying to say
24.   like (.) be happy (.)
25. EM: hm
26. GI: and just appreciate whatever skin colour you've got
27. EM: hm
28. GI: or skin tone (.) there's no point trying to change it I
29.   mean
30. ((approximately 4 minutes omitted))
31. EM: so what happened on Twitter?
32. (.)
33. GI: so it's like they had this whole light skin versus dark skin
34.   thing.
35. (.)
36. EM: oh really?
37. (.)
38. GI: yeah and it's like oh light skins are better than dark skin,
39.   (.) light skin people- (.) it's like (.) they say light-

40. EM: who started this?
41. (.)
42. GI: I don't know like you know people. (.) just random people.
43. ((approximately 4 minutes omitted))
44. EM: and then what else did you- (.) look the part that I thought
45.   was really interesting (.) is here. (.) Because I'm looking
46.   at linguistic stuff
47. GI: hm hm
48. EM: and here you changed voices?
49. GI: hm hm
50. EM: and then when you guys do your play (.) you know your
51.   sketches?
52. (.)
53. EM: like ((laughing) who are you being?)
54. (.)
55. GI: ((laughing) somebody else)
56. EM: somebody else?
57. GI: well that was actually real (.) that happened.
58. (.)
59. EM: it happened?
60. GI: so it's someone's voice (.) someone else's voice who- that
61.   said that to me.
62. (.)
63. EM: somebody said that to you?
64. (.)
65. GI: yeah. (.) well a few people actually.

Gina begins by explaining the overall meaning of the poem to Emilee. In lines 12–16, she refers to one of the events mentioned in the poem, being the discovery of lightening cream in her mother's bag. In line 17, Emilee searches for more information about the event, wanting to know when it had happened. Gina, in the following line, does not answer Emilee's question, but rather acknowledges it with a 'yeah', before continuing to explain the meaning, rather than the details, of the event, until line 29.

In the next part of the conversation, from lines 31–42, Emilee returns to a point made by Gina in the talk omitted in the fragment for reasons of space, being that the discourse around dark and light skin circulates through Twitter and other forms of social media. In line 40, she insists on gaining more detail about this, asking Gina who had initiated the Twitter thread. Gina responds in line 42 by telling Emilee that she cannot identify the circulating discourse with a particular instigator.

In the final extract of the conversation, from line 44, Emilee draws Gina's attention to the lines of the poem where Gina uses a different speech variety (as she also does at other times in the writing workshops, e.g. in performing sketches). In line 53, she asks Gina to identify whose voice she is using. Gina tells her that although the incident is real, it is not isolated. Rather, the voice indexes a recurring episode involving different people.

The extracts thus reveal a tension between anchoring particular lines and poetic resources used in the poem to specific times, places and people, and using those same devices for telling a bigger story. Through her conversations with Gina, Emilee came to understand that beyond narrating bounded incidents, Gina's poem employed those events to index how racism operates at larger scales. Gina's poem created an epistemological space in which Emilee began to understand how 'Whiteness' is an ideology that permeates education systems, is reproduced through different forms of media, and that supports economic, social and cultural dominance of White people (Gillborn, 2008). In this way, our conversations around Gina's poem allowed insight into broader issues of power and social injustice.

As a Matter of Closing

This chapter aimed to sketch out how a translanguaging approach, which involved being open to opportunities for transforming subjectivities, as well as social and educational structures (García & Li, 2014), panned out in one particular research process in a context of linguistic and cultural diversity. We have presented an exploratory description of how our collaborative ethnography has developed, and of some of the learning that this has afforded us.

As we mentioned in the opening of this chapter, in working together, we have sought more mutually beneficial ways of telling a story than is perhaps typical of academia. Negotiating other ways of working as university-based teacher-educator I researcher and poet I researcher brings with it different challenges – among them, how to fit more collaborative and 'poetic' types of research into the academic structures that we have. To give an example, in research such as we are doing, participant anonymity as usually required by ethics boards would not make any sense at all and would only serve to silence Gina. Rather, publishing Gina's poems as part of a scholarly volume and giving her full credit for them in doing so, is a way of sharing the benefits of her research participation more fairly. The preceding sections have shown how, besides the creative merit of Gina's work, her poem is also a material point of reference around which our shared research practice was able to develop.

As we do not have definitive conclusions to offer about this ongoing collaboration, we instead choose to end our contribution with a second poem in which Gina summarises, in her way, some of the themes we have discussed in this chapter. In the poem, she incorporates ideas and notions that circulated in our conversations, as well as others that form part of her knowledge repertoire (e.g. her learning about Erving Goffman's work through her A Level Sociology studies). Taking our collaborative work forward, a co-written poem to represent the research would be our next challenge.

Two minds
A poem by Ginalda Tavares Manuel

Crossing cultures
Crossing roads
Crossing styles
Crossing minds
No limits and no boundaries

When research collides with creativity and becomes one
There is ability for unity
Instead of opposition

Because poetry is like a piece of qualitative data
Holding quality, depth, reality

Some researchers tend sit on the opposite side of the glass wall
Than side by side in co-collaboration
The subject stays a subject
Nameless, anonymous

Unethically

Erving Goffman, a prime example of who got to know
Instead of just assuming
He studied the true person inside
Allowing the participant to participate

Creating validity, originality

Truth.

Understandably, there are reasons you keep people a secret
For ethical reasons
So not everybody is to blame

But who really holds the story-telling rights
Of one's life?

It's sometimes acceptable to:

Co-read
Co-write
And co-interpret

A researcher is a poet
And a poet is researcher
As the two are trying to find answers for life's questions

It's a sense of interculturalism and interconnectedness
But don't take my voice and lip-sync my words
Rather tell the world who this voice belongs to

Look deeper into building a shared understanding
Familiarity, a rapport that isn't just straight faced and
unemotional

But creates a connection to your research or researchee

Make your research come alive and speak truths across the orb
because you are always an inspiration to somebody reading

Notes

(1) Following Alexakos (2015), we use the symbol | in order to give equal importance to each of our dual roles.
(2) The following conventions are used in the chapter:
   Intonation:
      a. Falling: .
      b. Rising: ?
      c. Maintained: no symbol
   Pauses:
      a. Less than half a second: (.)
      b. Between half and one second: (..)
   Overlapping: [text]
             [overlap]
   Interruption: text-
   Lengthening of a syllable: te:xt
   Volume: °soft°
   Transcriber's comments: ((comment)) or ((comment) affected fragment)

References

Akkerman, S. and Bakker, A. (2011) Boundary crossing and boundary objects. *Review of Educational Research* 81 (2), 132–169.
Alexakos, K. (2015) *Being a Teacher | Researcher: A Primer on Doing Authentic Inquiry Research on Teaching and Learning.* Rotterdam/Boston/Taipei: Sense Publishers.
Bakhtin, M (1984) *Problems of Dostoevsky's Poetics.* Edited and translated by Caryl Emerson. Minneapolis, MN: University of Minnesota Press.

Blackledge, A. and Creese, A. (2017) Translanguaging and the body. *International Journal of Multilingualism* 14 (3), 250–268.

Bradley, J. and Moore, E. (2018) Resemiotization and creative production: Extending the translanguaging lens. In A. Sherris and E. Adami (eds) *Making Signs, Translanguaging Ethnographies: Exploring Urban, Rural and Educational Spaces* (pp. 91–111). Bristol: Multilingual Matters.

Budach, G., Kell, C. and Patricks, D. (2015) Introduction: Objects and language in trans-contextual communication. *Social Semiotics* 25 (4), 387–400.

Erickson, F. (1982) Classroom discourse as improvisation: Relationships between academic task structure and social participation structure in lessons. In L.C. Wilkinson (ed.) *Communicating in the Classroom* (pp. 153–181). New York: Academic Press.

Falthin, A. (2013) Transcription bank: Annika Falthin. *MODE: Multimodal Methodologies*, blog post, 1 May 2013, https://mode.ioe.ac.uk/2013/05/01/transcription-bank-annika-falthin/ (accessed 15 November 2018).

García, O. and Li, W. (2014) *Translanguaging: Language, Bilingualism and Education.* New York: Palgrave Macmillan.

Gillborn, D. (2008) *Racism and Education: Coincidence or Conspiracy?* Abingdon, Oxon: Routledge.

Goffman, A. (2014) *On the Run: Fugitive Life in an American City.* Chicago: University of Chicago Press.

Goodwin, C. (2000) Action and embodiment within situated human interaction. *Journal of Pragmatics* 32, 1489–1522.

Ibrahiim, K. (2016) The power of words. *Independent Leeds Magazine* 2, 11–13.

Jefferson, G. (2004) Glossary of transcript symbols with an introduction. In G.H. Lerner (ed.) *Conversation Analysis: Studies from the First Generation* (pp. 13–23). Philadelphia: John Benjamins.

Lassiter, L. E. (2005) *The Chicago Guide to Collaborative Ethnography.* Chicago: University of Chicago Press.

Latour, B. (2005) *Reassembling the Social: An Introduction to Actor-Network-Theory.* Oxford: Oxford University Press.

Lewis-Kraus, G. (2016) The trials of Alice Goffman. *The New York Times Magazine*, magazine article, 12 January 2016, www.nytimes.com/2016/01/17/magazine/the-trials-of-alice-goffman.html (accessed 15 October 2016).

Li, W. (2017) Translanguaging as a practical theory of language. *Applied Linguistics* 39 (1), 9–30.

McKay, S. and Bradley, J. (2016) How does arts practice engage with narratives of migration from refugees? Lessons from utopia. *Journal of Arts and Communities* 8 (1–2), 31–46.

Mondada, L. (2016) Conventions for Multimodal Transcription, pdf, July 2016, https://franzoesistik.philhist.unibas.ch/fileadmin/user_upload/franzoesistik/mondada_multimodal_conventions.pdf (accessed October 2016).

Moore, E. and Bradley, J. (2020) Resemiotisation from page to stage: The trajectory of a musilingual youth's poem. *International Journal of Bilingual Education and Bilingualism* 23 (1), 49–64.

Norris, S. (2004) Multimodal discourse analysis: A conceptual framework. In P. Levine and R. Scollon (eds) *Discourse and Technology: Multimodal Discourse Analysis* (pp. 101–115). Washington, DC: Georgetown University Press.

Pahl, K. (2014) 'It's about living your life': Family time and school time as a resource for meaning making in homes, schools and communities. In C. Compton-Lilly and E. Halverson (eds) *Time and Space in Literacy Research* (pp. 47–62). New York and London: Routledge.

Shah, A. (2017) Ethnography? Participant observation, a potentially revolutionary praxis. *Hau: Journal of Ethnographic Theory* 7 (1), 45–59.

Shuman, A. (1986) *Storytelling Rights: The Uses of Oral and Written Texts by Urban Adolescents*. Cambridge: Cambridge University Press.

Star, S.L. and Griesemer, J.R. (1989) Institutional ecology, translations, and boundary objects: Amateurs and professionals in Berkeley's museum of vertebrate zoology. *Social Studies of Science* 19 (3), 387–420.

Van Leeuwen, T. (1999) *Speech, Music, Sound*. Basingstoke: Macmillan.

Van Maanen, J. (1995) *Representation in Ethnography*. Sage.

Yanofsky, D., van Driel, B. and Kass, J. (1999) 'Spoken Word' and 'Poetry Slams': The voice of youth today. *European Journal of Intercultural studies* 10 (3), 339–342.

Part 3: Collaborative Outcomes

Comment on Part 3: Collaborative Outcomes

Zhu Hua and Li Wei

In talking about research and professional practice, outcomes are often presented in contrast with processes or aims and objectives. There seem to be assumptions that outcomes are tangible, specific and measurable, and research or practice is a linear process – one starts with aims and objectives, goes through processes and finishes with outcomes. The emergence of translanguaging not as a descriptive label, but as a research paradigm, challenges the established approaches to sociolinguistic research on multilingual practices by taking the research process itself as a social practice, and by setting the objective of transforming practice through doing research (Li, 2018; Li & García, 2017). The chapters in this section of the volume show that research is a collaborative, creative and critical process, and that the outcomes cannot be divorced from the process itself. Indeed, the research outcomes need to be conceptualised collaboratively, creatively and critically by all parties involved in the research process and in the context where the research is taking place.

There are three key senses of the *trans-* prefix of translanguaging (García & Li, 2014: 3):

(1) It refers to *a trans-system* and *trans-spaces*: that is, to fluid practices that not only go between, but more importantly beyond socially constructed, named language and institutional systems, structures and practices to engage diverse participants' multiple sense- and meaning-making systems and subjectivities;

(2) It refers to its *trans-formative nature*: that is, as new configurations of language and professional practices are generated, old understandings and structures are released, thus transforming not only subjectivities but also cognitive and social systems. In so doing, orders of discourses shift and the voices of Others come to the forefront, relating then translanguaging to creativity and criticality. Significantly, such transformations are done collaboratively involving all parties concerned;

(3) It refers to the *trans-disciplinary consequences* of the analytical approach, providing a tool for understanding not only language practices on the one hand and specific institutional structures on the other, but also human sociality, human cognition and learning, social relations and societal systems. In the meantime, the *-ing* suffix focuses our analytical attention to the dynamic nature of the practices including the research practice itself.

We, along with many other scholars, have argued that the notion of translanguaging represents a new way of thinking about how people communicate with each other, how they make meaning while making sense of what is said, or how they language (as a verb) (Bradley *et al.*, this volume; García & Li, 2014; Zhu *et al.*, 2017). Fundamentally, translanguaging is a process of knowledge construction as intended by the originators of the concept (Baker, 2001; Williams, 1994). This process is a creative, critical and collaborative one. Creativity and criticality are integral to translanguaging. For Li Wei (2011: 1223-1224):

> ...creativity can be defined as the ability to choose between following and flouting the rules and norms of behaviour, including the use of language. It is about pushing and breaking the boundaries between the old and the new, the conventional and the original, and the acceptable and the challenging. Criticality refers to the ability to use available evidence appropriately, systematically and insightfully to inform considered views of cultural, social and linguistic phenomena, to question and problematize received wisdom, and to express views adequately through reasoned responses to situations. These two concepts are intrinsically linked: one cannot push or break boundaries without being critical; and the best expression of one's criticality is one's creativity. Multilingualism by the very nature of the phenomenon is a rich source of creativity and criticality, as it entails tension, conflict, competition, difference, change in a number of spheres, ranging from ideologies, policies and practices to historical and current contexts. ... the enhanced contacts between people of diverse backgrounds and traditions provide new opportunities for innovation, entrepreneurship, and creativity. Individuals are capable of responding to the historical and present conditions critically. They consciously construct and constantly modify their socio-cultural identities and values through social practices such as translanguaging.

The third dimension – collaboration – is a crucial epistemological stance towards (translanguaging) research and practice. As a pedagogical practice, the idea of collaboration is not new to multilingual education research where collaboration in the form of co-participation and co-construction are often discussed with a focus on equitable access to resources, equal contributions from individuals and emergence of new knowledge through learning. Li Wei (2014) uses the notion of 'co-learning' to emphasize the benefits of breaking down the dichotomy of the teacher/instructor and the learner and to highlight what they can all gain from the collaborative process of knowledge construction. Applying the same principles to research practice, a collaborative stance is about reconceptualising the ownership of the research, repositioning the relationship between the researcher and researched and transforming (sometime hidden) hierarchies and power dynamics between the researcher and the researched, and regarding them all as participants of the research process. The four chapters in this part of the volume all engage with these three dimensions of creativity, criticality or collaboration.

There is a lot to unpick in Lou Harvey's thought-provoking account of a collaborative creative inquiry. We see a chain of 'voices' (in Bakhtin's [1981] terms) and creative ideas travel from one site to another and bounce off from one person to another. Raj's story of culture/language shock and his feelings of 'discomfort' in his first few months of being an international student in the UK, told in his own voice in an interview (positioned as a talk 'with' the researcher), became stimuli for creative activities as part of the process of producing an artistic performance. The subsequent stages of ideas generation and sharing among the team developed these artistic possibilities into a seven-minute performance which runs two parallel narratives, with one narrative involving the audience's participation. The audience were then invited to write down their reactions to the performance. It is interesting to see that about half of the descriptive words given by the audience are related to negative emotions, including the feeling of discomfort and disruption, the very feeling conveyed by Raj in his narrative. Thus we see a full circle of the re-semiotisation of voices and ideas in the creation of the performance. This is a brilliant example that brings to the fore the complexity of the issues of whose voice, whose authorship, and ultimately, who are collaborating/co-producing in artistic performance/research. To what extent does Raj's contribution to the performance count as collaborative? Harvey terms these issues eloquently as a phenomenon of 'entangled trans-ing', in which translanguaging, transcreation and transauthorship are brought together in the making and delivery of the performance. Harvey's creative, critical and collaborative inquiry lends support to the argument that process and outcome do not exist in dichotomy.

Kendall A. King and Matha Bigelow's chapter brings us to classrooms in Minnesota where the researchers make collaborative efforts with students, teachers, and administrators, among others, to 'normalise' translanguaging pedagogies locally. The hurdles to their efforts are not hard, visible, legal barriers, because multilingualism has been recognised as a resource as part of progressive language education policy introduced by the state of Minnesota in 2014. Rather, the real challenges are from the mindset which takes translanguaging pedagogy as something novel, experimental and therefore, 'impossible', and from the reality check identified as the lack of institutional support, bilingual teachers or curriculum, as well as diverse language ranges and the mobile nature of the families concerned. How to resolve these issues? We find some answers in the four examples provided in the chapter. The first two examples illustrate how creative and critical translanguaging practices could give life to classrooms, generate new energies, and afford participatory and collaborative opportunities. The teaching materials for choral reading activities, *Sam and Pat*, which could not be further removed from multilingual students of refugee backgrounds, were replaced with the materials developed collaboratively among students, drawing from their own knowledge about their families and Somali cultural practices, and constructed through the Somali language initially and translated into English. The introduction of multilingual and multi-modal discussion on Facebook creates a space for English learners of refugee backgrounds to use all of their linguistic, cultural and digital resources to engage in making sense of the world around them. Despite these initial successes, however, the researchers note reflectively that translanguaging pedagogy did not get sustained or adopted by other schools in both cases. What is needed along with these on-the-ground curriculum innovations are policy initiatives that would help to systematically recognise the status of 'other' languages of English learners. One of them is the development and implementation of placement assessments in students' first languages and the other is probing the uneven uptakes of official recognition of competences in languages other than English in the form of 'multilingual seals and certificates' across languages and areas. These initiatives bring policymakers and institutional key players into the expanding list of collaborators – translanguaging pedagogies are not just a matter of 'the skill or the will' from those on the ground, but a campaign of public awareness and advocacy.

Moving on to linguistically and culturally diverse Barcelona, Júlia Llompart-Esbert and Luci Nussbaum's chapter puts forward a two-part argument, i.e. (a) students are researchers themselves; and (b) translanguaging pedagogy needs to place students as the key drivers. Their argument not only legitimises students' involvement in research about educational practice which concerns their very own learning and experience and restores agency and responsibility among students, but also breaks down the traditional boundaries between the researcher and the researched. It liberates students from their default positions in the research process – i.e. passive, somewhat willing and cooperative in some cases, and less forthcoming in others, recipients of intervention, targets of observation and producers of 'data'. So what can we do to develop students as researchers and to enable creative and critical translanguaging practices? The authors provide examples illustrating the journey these students travel from a monolingual ideology to growing recognition of flexible multilingual practices. In these examples, we see evidence of creative and critical translanguaging practices when students compared, and subsequently reflected on, what is going on in real life and what they thought or took for granted when they were tasked with documenting language uses through field work; when they collaboratively constructed the meta-commentaries about the slippery boundaries between Spanish and Catalan and inadvertently developed the conversation into a banter; and when they gave a presentation about multilingual environments with their nuanced observations.

The last chapter by Claudia Vallejo Rubinstein adds another perspective to the cycle of translanguaging outcomes. The research team gained rich insights towards everyday translanguaging practices of children such as a 10-year old girl of a migrant family background, Hasu, and her peers who attended an after-school literacy programme in a multicultural, multilingual primary school in Barcelona. While Hasu, new to the region, appeared to struggle with literacy programmes at school, she demonstrated an extraordinary ability to call upon her knowledge of Punjabi, English, Catalan and Spanish and all her other available resources to generate meaning and make herself understood. These cases of translanguaging practices were then shared with, and analysed and interpreted by pre-service teachers as training materials, in their preparation for the task of developing teaching activities and materials that aim to facilitate the implementation of a translanguaging pedagogy. The developed activities and materials were then brought back to the after-school clubs and became part of the resources available for the children in every session. The project is an exemplary endeavour of developing and implementing translanguaging pedagogy creatively, critically and collaboratively.

The four chapters in the section together make a strong case for seeing translanguaging outcomes as a collaborative, creative and critical process, underpinned by strong commitment to social justice and equality in the translanguaging approach to research and practice. Along with the chapters in the volume as a whole, they open up new venues for future research and extend the translanguaging approach to wider contexts. They invite the reader to consider questions such as:

- Who are the key stakeholders and drivers in research and practice?
- Who are the researchers? Who are the researched? Who owns and shares the outcomes?

- Is everyone equal in the process of collaboration? Who cooperates with whom?
- How can de facto hierarchy and power dynamics be transformed through the research process?
- How should we measure successes of collaborative research and practice?

These are some of the key questions facing our collective endeavours in rethinking research through the translanguaging lens. The 'trans-' problematises the status quo in the relationship between the researcher and the researched, takes the researcher out of their comfort zone and destabilises their often taken-for-granted privilege and associated authority. In collaborative research, many issues remain to be debated and written about. For example, what do we do when the participants have their own agenda that they would like to achieve through the research process or through association with the research? In our own research, we had participants who were in precarious or sensitive positions in society and wanted to use our status as academics and affiliations in pursuit of their socio-political (and sometimes controversial) causes. We also had a case in which the key participant, well-educated and well-versed in the ins and outs of research, played the participatory game. They readily produced the buzz terms, terms that they thought we would like to hear, in the interview. How does the researcher differentiate the constructed reality from performance and to what extent do we need to do so? These are further questions that we need to address if we continue to pursue the 3Cs, i.e. criticality, creativity and collaboration, from the translanguaging research perspective.

References

Baker, C. (2001) *Foundations of Bilingual Education and Bilingualism* (3rd edn). Clevedon: Multilingual Matters.

Bakhtin, M.M. (1981) *The Dialogic Imagination: Four Essays* (M. Holquist, ed.; C. Emerson and M. Holquist, trans.). Austin: University of Texas Press.

García, O. and Li, W. (2014) *Translanguaging: Language, Bilingualism and Education.* London: Palgrave Macmillan.

Li, W. (2011) Moment analysis and translanguaging space: Discursive construction of identities by multilingual Chinese youth in Britain. *Journal of Pragmatics* 43 (5), 1222–1235.

Li, W. (2014) Who's teaching whom? Co-learning in multilingual classrooms. In S. May (ed.) *The Multilingual Turn* (pp. 177–200). London: Routledge.

Li, W. (2018) Translanguaging as a practical theory of language. *Applied Linguistics* 39 (1), 9–30.

Li, W. and García, O. (2017) From researching translanguaging to translanguaging research. In K.A. King., Y.J. Lai and S. May (eds) *Research Methods in Language and Education* (3rd edn) (pp. 227–240). Berlin/Heidelberg/New York: Springer.

Williams, C. (1994) Arfarniad o ddulliau dysgu ac addysgu yng nghyd-destun addysg uwchradd ddwyieithog [An evaluation of teaching and learning methods in the context of bilingual secondary education]. PhD thesis, University of Wales.

Zhu, H., Li, W. and Lyons, A. (2017) Polish shop(ping) as translanguaging space. *Social Semiotics* 27 (4), 411–433.

9 Entangled Trans-ing: Co-Creating a Performance of Language and Intercultural Research

Lou Harvey

Introduction

I am an education researcher with a focus on language and intercultural learning in higher education contexts. This chapter offers an account of the co-production of a piece of performance art with theatre company Cap-a-Pie, based on my research into the English-language learning of six UK-international students. I first outline the context of this research and demonstrate how the concept of *voice*, as understood by Mikhail Bakhtin, was fundamental to my theoretical and methodological frameworks. Following this, I describe the collaborative creative inquiry activities we engaged in as part of the making process, and the resulting short performance. I return to the concept of voice for my analysis, demonstrating how voice was the material vehicle for different yet entangled processes of *trans-ing* (following Mylona, 2016) – *translanguaging, transcreation,* and *transauthorship* – in the making and the performance. I conclude that this analysis offers potential for enhancing the transformative possibilities of a translanguaging approach by accounting for the material and moral dimensions of language education.

Research Context

The research on which the performance co-produced with Cap-a-Pie was based was a narrative study of six UK-based international university students, and their motivation for learning English across their lives (Harvey, 2014, 2016, 2017). The following analytical vignette, based on Harvey (2014), summarises the story of one of the participants:

Raj [aged 19–20 at the time of the research] was a very successful English learner in India, and was highly proficient in English when he came to the UK at the age of 16 to go to boarding school, his first experience of being immersed in a fully English-speaking environment. The move was a substantial language shock, and he found the local accent and variety of English very different to the English he had used in India. This shock made him unable to speak for the first month or two after his arrival – he would just nod silently in response to anyone who spoke to him, feeling like he was living in a prison inside my mouth'. After listening and becoming adept at the local English, Raj was able to 'pass' as British, to the extent that some of his schoolmates could not believe that he was actually Indian. However, he began to feel that this local English was not right for him; he wanted a voice through which he could express both his Indian/international identity and his aspiring British identity, a voice which he had chosen and with which he could be recognised on his own terms. This involved learning to negotiate the perceptions of England he was brought up with, as a highly cultured civilisation and a pinnacle of civil values, and his own perceptions of English life as a resident, and choosing which values and aspects of lifestyle he wanted to adopt. He developed a hybrid accent which he felt carried enough of the prestige of a British accent and through which he could also be recognised as Indian, which would allow him to fit into both contexts. In this way, he understood himself to be becoming 'in some ways British, in some ways Indian', and able to 'fit in here while you are here ... fit in there while you are there'.

Central to my analysis of the participants' stories was the concept of *voice*. In Bakhtin's dialogic perspective (Bakhtin, 1981), using utterances, whether in speech or in thought, always involves using the utterances of others, and thus both outward and inner speech are social phenomena; consciousness itself always exists in relationship with other consciousnesses, dialogically constituted through the continuous dynamics of communication. This 'dialogical self' is also an embodied self, with biological and biographical unity deriving from its specific socio-historical conditions. Each embodied self carries an individual voice, with a unique and distinct *emotional-volitional tone* (Bakhtin, 1993), through which individuals express themselves as sociohistorically specific people located in a particular time and place, with their particular ways of communicating, being and knowing. Through the development of voice and emotional-volitional tone, speakers author themselves in response to other voices in the world around them, putting their signature to their utterances, and their own accent on linguistic and communicative forms which have been used many times before (see Hicks, 2000: 240). This development is an ongoing and dynamic process of embodied engagement with individual voices, and with ideological, historical and social forces; like consciousness, it is both shared and social, and uniquely individual and embodied. This is the *ideological becoming* through which I was able to interpret Raj's, and the other participants', motivation for learning English, and becoming English speakers, across their lives. A Bakhtinian understanding of *voice* might therefore be summarised as

(a) voice as utterance – the ability to employ, and be understood in, recognised communicative modes, through which we perform
(b) narration – a material enactment of our own embodied biological and biographical self, which becomes an act of
(c) authorship – a response to, and positioning of oneself among, other individuals and social forces.

The concept of *voice* was also central to my methodological approach and research design. In order to give the participants the opportunity to reflect on and theorise, as well as relate, their own experience, I designed a dialogic interview methodology which created theoretical space for us to talk *with* each other (following Bakhtin, 1984: 68) rather than me talking *to* them, while also recognising that I needed to acknowledge my researcher role and my authorial position. This involved using my own analytical interpretations of the participants' stories as the basis for further interviews, then analysing the six stories together and sharing this collective analysis with each participant, so that they were responding to elements of the others' stories as well as their own. In this way, I explicitly shared the interpretive process with the participants, and the analysis became part of their lived experience, part of their stories (see Harvey, 2015). To theorise this, I drew on Bakhtin's concept of the *polyphonic* novel, in which the author creates '*freedom for others' points of view to reveal themselves*' while maintaining a '*positive and active quality*', in which the characters are not objectified but rather re-created 'in their authentic *unfinalizability*' (Bakhtin, 1984: 67, 68, italics in original). The study therefore wove together the 'living dialogic threads' (Bakhtin, 1981: 276) of our voices: mine and theirs. In this chapter, I will demonstrate how a Bakhtinian understanding of voice is applicable to the making and the performance of our piece, *Up and up and up towards*, and how it might usefully inform and expand understanding of translanguaging as transformation. Specifically, I suggest that a Bakhtinian perspective extends the semiotic understanding of communication on which translanguaging is currently based, and which only accounts for voice as *utterance*, towards an understanding of the material and moral dimensions of intercultural communication and language education by accounting for voice as *narration* and as *authorship*.

The Making Process: Creative Inquiry

In June 2015 I took part in the Leeds Creative Lab, a space for facilitating collaboration between three arts groups and three academics, in which I was paired with theatre company Cap-a-Pie – Artistic Director Brad McCormick, and Producer Katy Vanden. Our brief was to spend three days together over June, and then come to a showcase event in early July where we would talk to the rest of the Lab about what we had done. There was no requirement for a product or an output, but when we came together at the initial meeting we decided we would like to try and make a short performance for the showcase. (See Harvey, 2018 for discussion of the performance as a public engagement project).

Our starting point was for Cap-a-Pie to read the research stories I had created for the thesis (with the participants' permission). I then spent three days at their office in Newcastle, where we began by talking about my research, where it had come from, the ideas in it, Cap-a-Pie's responses to the stories, and why I wanted to adapt it for performance. This process of open discussion was very useful because it forced me to articulate the crux of the idea I wanted to communicate: the assumption that you can assume you know what is going on around you, but then you realise that actually, you do not, and that there are unspoken rules you are not aware of. The key idea was therefore about not understanding, especially in a situation when one expects to be able to understand.

This concept gave us our first idea for the form of the piece. Brad suggested that we could use this idea to apply to the theatre too, where the semiotics of the theatre could represent the semiotics of communication, and by playing with the audience's expectations around these we could put them in the position of discomfort that my research participants had felt. We realised that we could use the theatrical/performance context to not only play with people's expectations, but to render the process of performance explicit by 'making the mechanics visible', to use Brad's phrase. We considered the various ways in which we could do this, such as asking the audience to participate; switching between stories without linking them; the actor stepping out of character; using other languages, silence or loud music without apparent purpose or warning; and inconsistent use of gesture or changing the gestures associated with different characters.

We then moved from thinking about form to thinking about content. To generate ideas for this, we did Cap-a-Pie's version of creative inquiry, a method that mixes creative activities with the precepts of Philosophy for Children (see www.philosophy4children. co.uk). It involved starting with a stimulus, which for us was a piece of verbatim text from participant Raj's story. This was the text we used from this story, a section which had particularly resonated with Brad:

> yeah I mean the main thing I suppose in explaining a situation like that is choosing whether I could fit if I wanted to or not... there's this whole thing you know about people just being sort of shunned out of a community because of the way they look and sound... it's pretty hard to fit in... I think I had a breakdown after like at the end of one month... I was just in my room full on crying and that's when I decided that no you can't sort of let them choose whether you fit in or not... and it's much more important that you choose whether you can fit in or not... when I did open my mouth ultimately after the one or two months it was always to you know... I would say not necessarily imitate but sort of in an attempt to sort of settle in... and I sort of really would like to contrast that with... choosing to completely take up their lifestyle you know?... and sort of say it's the basic difference between whether you can fit into a social group and whether you can sort of completely take on the lifestyle

Brad read this extract aloud twice, and then we did a free-association activity where we called out themes related to what we had just heard. The themes could not be the same as actual words in the text, but could be anything else. Our list was:

other	away-ness (from home)
socialising	voice
acceptability	freedom
comfort	ambition
silence	determination
potential	copying
the senses	inclusion
struggle	

We then did a series of activities in which we responded creatively to these themes in different ways. These were done quickly and roughly, within short time limits – they were not meant to be polished or well thought-out, but to be raw and related to our immediate associations with the themes. Although it is necessarily superficial, there is not space to present all of these here and so I will present only a selection of those activities which ended up having the greatest relevance for the final performance. We wrote a concept for a business, a concept for a restaurant, memories, and imagined dialogues, each based on two themes of our choice. One of mine was a concept for a business called 'The Hangout', based on the themes of *comfort* and *socialising*:

> where customers pay to come and sit and chat + make friends. Different types of space, e.g. for debate (w/chairs and tables facing each other), w/topics to choose from, all the way up to comfy chairs and coffee for relaxed chat. Employees paid to engage in conversations

Brad wrote a concept for a business, 'Culture Class', based on *socialising* and *other*:

> This is a service for people with xenophobia who feel it is something they need to improve. The company offers meet-and-greet nights where everyone else there is guaranteed to be from another country. There are games, conversation starters

I wrote a memory of 'something which happened away from home', based on *ambition* and *the senses*:

> June 2006, Bratislava, in the Smíchovská Perla pub – Rachel, Phil, Simon, Milan, me. The pub had these rubber mats all over the floor and was green and brown and smelt of beer – it was comforting and reminded me of my Dad's old cricket clubhouse. The World Cup [football] was on – England v Germany. Phil and Simon were talking disparagingly about someone, saying 'oh well she's very ambitious', and I said 'what's wrong with being ambitious? I'm ambitious' and they said 'oh well, better make sure you don't walk all over everyone on your way up then'. It was part of Rachel's and my brilliant day and when she and I talk about this day now, we remember this conversation.

Brad wrote an imagined dialogue based on *socialising* and *comfort*:

> L: why did you have to take me here?
> A: because you need to get out, be amongst people.
> L: this is too many people.
> A: but there are better odds of meeting them, it's statistics.
> L: I feel a bit faint.
> A: everything I try, you don't like.
> L: that's not true, it's just this one is especially difficult for me.
> A: for Christ's sake!
> L: what?

A: excuses get people nowhere, get rid of the excuses and you'll move forward. simple as that.

After writing the dialogues, we swapped our papers and asked each other to consider: who are these characters? What are their backgrounds? Brad passed his paper to me, and I wrote that the characters were Liesl, a 22-year-old Austrian woman living in Newcastle, fairly new to the UK, and Andy, her British boyfriend. We then swapped again, and I passed the paper to Katy to answer the question: what's the context of the conversation? Katy wrote

- Liesl meets Andy at an interesting social event where participants are given a series of interactive activities [referencing Brad's 'Culture Class']
- Andy and Liesl start dating but Liesl didn't like meeting Andy's friends
- Liesl and Andy break up as Andy thinks she's making no effort to make friends and he doesn't want to take care of her anymore

Our final activity of the day was to each come up with two philosophical questions, each of which should incorporate two of the themes. This needed to be an open question which had some meaning to us and which we would want to discuss. We then read our questions aloud and voted on the one we liked best, which was

- What ambitions are acceptable?

We all felt that this connected well with the research, and that it would enable artistic possibilities. This question became a kind of conceptual basis for the performance.

Between the first and the second day, Brad came up with the idea of using two parallel narratives: one that tells a well-known story, for which we chose the story of Icarus, the Greek myth about the boy whose ambition became his own destruction when he flew too close to the sun with his wax wings; and one telling the story of Liesl and Andy. In an open discussion we talked through possibilities for developing Liesl and Andy's story. Bearing in mind our philosophical question, and thinking through ideas of ownership of the language, especially as these arose from Raj's story, we decided that the characters would argue because Liesl likes to use local English and local Newcastle dialect words, but Andy feels she should not be speaking this way because she is not a local British person. These two stories therefore became the content of our seven-minute performance, which we called *Up and up and up towards*. The performance, by Brad, ran as follows:

Brad starts by narrating the story of Daedalus and Icarus. He physically enacts the characters to identify them (e.g. placing a crown on his head to symbolise the king, using his fingers as two horns for the Minotaur, working with a chisel for Daedalus). He steps out of character to comment on his performance and the story (e.g. 'I think that's the wrong gesture' or 'All that human sacrifice seems like overkill!'). He starts to confuse the gestures, performing Daedalus with a crown on his head, the Minotaur with long flowing hair, and so on. At one point he stops speaking for a minute in the middle of a sentence; at other points he speaks gibberish for sections of the script. At the climax of the Daedalus and Icarus story, the narrative shifts without warning: 'Daedalus watched with horror as his son flew up and up and up towards the pub that Andy and Liesl go to on Friday nights. Now everyone act like you would in the pub on Friday night'. Brad then acts out a conversation between Andy and Liesl, taking on the role of Andy and asking different audience members in turn to play Liesl, telling them what to say (e.g. 'What would you like to drink? Now you say: a pint of lager, please'). The conversation, which becomes an argument, is about Andy's frustration with Liesl's adoption of the local dialect – he is embarrassed about her speaking Newcastle English because he feels she shouldn't be talking that way as a foreigner who already speaks English well. Brad/Andy continues to ask different audience members to play Liesl ('Why are you talking like that? Now you say: Because I like the sound of the words and I want to fit in'). For the final turn in the dialogue, Brad/Andy does not tell 'Liesl' what to say, leaving an embarrassed silence at the end of their argument and the audience member dangling. Brad then reverts to being the narrator, telling us that soon afterwards, Andy and Liesl split up and Liesl flies to London. Liesl's plane taking off is followed by a shift back to the original Daedalus and Icarus story: 'as she flew up and up and up towards the sun Daedalus shouted to Icarus No, Icarus, come back, don't go too close! … and saw the feathers floating past him through the air'. (adapted from Harvey 2018: 381–2)

We performed *Up and up and up towards* to colleagues and friends at the Creative Lab showcase event (July 2015), at another University of Leeds engagement event (January 2016), and to the general public at city-wide arts event Leeds Light Night (October 2016). We asked each audience to write down three words to describe the performance, and three words to describe what it was about. This is a composite of their responses (as there were many similar answers):

Description	What it was about
Uncomfortable	Non-verbal communication
Unpredictable	Interweaving stories
Multifaceted	Languages
Withdrawal	Silences
Anticlimax	Interruptions
Disruption	Fantasy
Incongruous	Not understanding
Immersive	Communications
Engaging	Filling in the gaps
Humorous	Storytelling
Delightful	Setting up and breaking convention
Constructive	Semiotics
Neutral	Making sense
Absurd	
Celebratory of narrative	
Suspenseful	
Spontaneous	

It is evident from these responses that the audiences understood what we were trying to do. The emotional responses were very varied, and included the confusion and discomfort we were hoping for (uncomfortable was the most common response). But there were also some unexpected responses, especially around humour – it had not occurred to us that it was funny, but it was funny precisely in the subversion of expectations, and in the embarrassment and confusion this generated.

Discussion

Through a trans-lens

To analyse the process of making the performance, I return to the concept of voice, and to consider the many voices which made the performance. Firstly, and perhaps most obviously, there was my own authorial voice as the researcher. However, the original research was explicitly dialogic, co-constructed not only between me and the research participants (who themselves speak in complex and hybrid voices, as Raj's vignette indicates) but among the research participants themselves, as they engaged with each other's voices (see Harvey, 2015). Cap-a-Pie then had their own engagement with the research and the participants' voices, as they read the stories and prepared the activities for the creative inquiry. Then Cap-a-Pie and I came together for the creative inquiry, weaving together the 'living dialogic threads' (Bakhtin, 1981: 276) of all our voices into a new creation. And this creation invited the audience to bring their voices, both physically in their participation as Liesl, and symbolically in terms of engaging their own sense-making processes to understand the performance. It is impossible to separate these voices in the making and performance of *Up and up and up towards* – they have all played a part for the performance to come into being. To analyse the different ways in which voice(s) work(s) in *Up and up and up towards* I draw on the concept of *trans-* as it is used by performance scholars Amelia Jones and colleagues, performing movement across, through and beyond.

Jones (2016) claims that when trans- is understood as process – a process of moving across, through, and beyond – it highlights 'our relationship to knowledge creation as performative ... the trans- is itself fluid and multipurpose, a mode of performing complex relationships between one site, identification or mode of speaking/ doing/being and another' (2016: 4, 2). The performance of complex relationships entails fluidity and porosity of boundaries, without effacing boundaries completely; trans- is implicitly relational and ongoing (as performed by the hyphen) (2016: 1). Both the artificiality and the transgression of boundaries is fundamental to the concept of translanguaging, and scholars adopting a translanguaging approach acknowledge the boundary work which takes place across, through and beyond named languages (see e.g. Blackledge & Creese, 2017; Bradley *et al.*, 2018), moving towards an understanding of translanguaging as 'extending beyond language, to encompass multiple modes of communication and interconnectedness' (Callaghan *et al.*, 2018: 50). In an extension of the translanguaging lens, Bradley and Moore have identified the ways in which meanings can be traced and transformed through movement across a range of semiotic resources and processes in their analysis of the *re-semiotisation* of a Slovenian folk story for performance as street theatre, and of a written poem for spoken/sung performance (2018, following Iedema, 2003). They point to the conceptual power of semiotic transformation, positing that linguistic methods alone are insufficient for understanding such transformation of meaning. This extended understanding of translanguaging is appropriate to the making and performance of *Up and up and up towards*, in which semiotic transformation took place in the movement across the oral, written and gestural modes. Bradley and Moore conclude by questioning 'the role of the linguistic repertoire as central in and paradigmatic to communication, and to translanguaging in particular' (2018: 109). In my own argument, I extend this question further and posit that in the creative inquiry and the performance, this semiotic understanding of communication only accounts for the first aspect of voice defined above: voice as *utterance*, or the ability to employ and be understood in recognised communicative modes. In order to more fully understand how the transformation of meaning took place in this context, it is necessary also to consider voice as *narration* and voice as *authorship*.

In an understanding of voice as narration, individuals express themselves as sociohistorically located, with their own ways of being and knowing. Ways of knowing are fundamental to Jones' trans-, which allows for 'an understanding of a field of knowledge in a momentary and processual way', which 'enables rather than confirms or fixes knowledge about the world' (Jones, 2016: 5, citing Gotman, 2016). The making and the performance of *Up and up and up towards* highlighted fluidity and process – it did not make claims for any individual's knowledge or attempt to reify this knowledge. We did not simply translate knowledge from one domain to another; we moved not only across and through semiotic modes, but also through the 'real' and the fictional, the literal and the symbolic, the well-known myth and the newly-created narrative, narrative disjuncture and discontinuity, the global language and the nonsensical gibberish. In this sense, the creative inquiry and the performance can be understood as processes of knowledge *transcreation* (Huang & Fay, 2016), a multi-directional and inter-transformative process of merging, encompassing and (re) creation – a trans-cultural, trans-disciplinary, trans-practitioner, trans-methodological and trans-conceptual space of thinking and interthinking (Huang & Fay, 2016, np). The creative inquiry moved beyond knowledge exchange, as we drew on our collective experiences and imaginations to create new stories. Similarly, the performance moved beyond dissemination of my research; it was not meant to tell people what I knew, as it was not 'my' knowledge. Rather, it enabled and facilitated knowledge in the moment. It invited sensing rather than thinking (Mylona, 2016) in the affective experience: the feeling of confusion and discomfort when the audience realised they did not know what was going on; the feeling of awkwardness in their embarrassed laughter. In this way, the performance was an enactment of multiple, inextricably connected narrative voices which enabled the audience's own knowing; an enactment of transcreated knowledges based on our unique experiences as embodied biological and biographical selves.

To consider how voice operates as authorship in the performance, I have drawn on the After Performance Working Group's (2016) process of *ensemble thinking* (see Andrews *et al.*, this volume, for a fuller analysis of collaborative thinking), which they characterise as 'a mode of transauthorship' (2016: 35) through which 'our voicing percolates' (2016: 35, 36). The 'percolating voices' here are redolent of Bakhtin's 'living dialogic threads' (Bakhtin, 1981: 276), the voices woven together as part of the original research, the creative inquiry, and the performance. The After Performance Working Group (2016) state that

> We use the prefix 'trans-' not to attempt a transcendence of the figure of the author or of authorship, but to attest to an

experience of transmutating beyond the boundaried authorial self, towards moments of transparency, transmission, and the transit – or more definitively the circulatory movements – of ideas. (2016: 36)

The concept of *transauthorship* therefore points to ideas moving out of traditional notions of ownership and into a process, a circulation, a series of moments of understanding. In *Up and up and up towards*, the complexity and inextricability of the different narrative voices make it impossible for any voice to claim belonging, and they open the performance up to interpretations beyond those associated with bounded authorial selves. Furthermore, the participatory elements invite the audience's authorship, as the performance cannot exist without them, and although they are told what to say, nevertheless their performances are always unpredictable: in one performance, a man played Liesl with a heavily affected Austrian accent; in another, a woman who had been told to ask for a pint of lager asked for a vodka and orange instead. The audience participation also enacted a kind of transgression of authorship, an enforcement of it which generated a crucial affective response in the audience: the feeling of anxiety that they were being called on to participate in this performance they did not quite understand; the disempowering feeling (but perhaps also a feeling of relief?) of being asked to repeat a sentence, like a language learner, that made sense in terms of the scenario but was not necessarily what they would choose to say themselves; the feeling of exposure (or betrayal? liberation?) when Andy did not give Liesl her final line. In this way, *Up and up and up towards* can be said to be a *transauthored* product in which the entwined voices enable multiple responsive positions for the audience, simultaneously enacting (the lack or loss of) voice as authorship and drawing attention to its transauthored nature. I could say, after Barthes, that the performance is 'a new object, that belongs to no-one' (Barthes, 1984: 72). However, as a transauthored product it is constantly made anew every time it is performed to a new audience, and therefore belongs to everyone - it cannot exist without any one of us.

Entangled trans-ing

The usefulness of the *trans-* lens in this analysis, then, is to draw attention to the performativity of knowledge creation, and to the inseparability of *what* is happening from *how* it is happening (see Bayley, 2016). The sense-making in *Up and up and up towards* went beyond semiotic modes; it made sense by *not making sense*. It offered the audience an experience of being unable to make meaning in a linguistically narratable way, as they would expect to be able to do – just as the original research participants expected to be able to understand when they came to the UK, and then found that they couldn't. This experience had to be made sense of affectively, and just as the research participants' learning took place through *being*, so were the *what* and the *how* of the audience's learning inseparable in the embodied, affective experience of not understanding. The modes of communication, the experiences and knowledges, and the authors, cannot be clearly identified; they are entangled (following Barad, 2007), moving across, through and beyond boundaries through the material vehicle of *voice*. To name them would miss the performative point: this would be a reversion to representationalism, using language to make agential 'cuts' which create boundaries, separating matter into subjects and objects (Bayley, 2016: 45, following Barad, 2007). However, through a *trans-* analysis, we find a way to both consider the different ways in which the material vehicle of voice operates, while also acknowledging and attending to the fluidity and porosity of boundaries. *Trans-* enables the tension of what Bakhtin describes as 'creative understanding', where each voice 'retains its own unity and open totality', but all are 'mutually enriched' through their engagement with the others (Bakhtin, 1986: 6–7). Our voices are dialogic, emanating from uniquely embodied beings yet always-already entangled with other voices. And in my analysis translanguaging, transcreation and transauthorship are themselves entangled in voices, offering different-yet-related *trans-* lenses which move across, through and beyond each other in the process of knowledge creation. This analysis of entangled trans-ing (following Mylona, 2016) therefore offers a new relationship between translanguaging and transformation: broadening the analysis beyond language and semiotics and into *voice* may shed a wider pool of light onto how knowledge is created, and how learning takes place, across, through, and beyond various boundaries.

Conclusion

This analysis of entangled trans-ing draws attention to the complex role of *voice* in communication and knowledge creation, and to the potential for 'extending the translanguaging lens' (Bradley & Moore, 2018) beyond sociolinguistic and social semiotic analysis. I suggest that extending the translanguaging lens to include consideration of narrative selves and authorship has the potential to make two important contributions; namely to understanding the material and the moral dimensions of communication and education. Emerging arts-based approaches to intercultural communication and language education (Bradley *et al.*, 2018; Frimberger *et al.*, 2017; Harvey,

McCormick and Vanden, 2019; Harvey *et al.*, 2019) engage with materiality and performativity in order to account for affective, embodied and multisensory ways of understanding and becoming (what Frimberger (2013) calls 'messy languaging') and to de-centre language as the primary sense-making mechanism, recognising that 'language has been granted too much power' (Barad, 2007: 132). There are also an increasing number of perspectives on the moral and ethical dimensions of language education in a world of communicative multiplicity and complexity (Clarke & Hennig, 2013; Harvey, 2016; Kubanyiova & Crookes, 2016), in which language and intercultural learning may be understood as processes of responsive and responsible authorship, of becoming in the world with others (Harvey, 2017). Further consideration of these dimensions will be crucial to an engaged and critical future understanding of translanguaging as transformation.

References

After Performance Working Group (2016) After performance. *Performance Research* 21 (5), 35–36.

Bakhtin, M.M. (1981) *The Dialogic Imagination: Four essays* (M. Holquist, ed; C. Emerson and M. Holquist, trans.). Austin: University of Texas Press.

Bakhtin, M.M. (1984) *Problems of Dostoevsky's Poetics* (C. Emerson, ed. and trans.). Minneapolis: University of Minnesota Press.

Bakhtin, M.M. (1986) *Speech Genres and Other Late Essays* (C. Emerson and M. Holquist, eds.; V. W. McGee, trans.). Austin, TX: University of Texas Press.

Bakhtin, M.M. (1993) *Toward a Philosophy of the Act* (V. Liapunov and M. Holquist, eds.; V. Liapunov, trans.). Austin, TX: University of Texas Press.

Barad, K. (2007) *Meeting the Universe Halfway: Quantum Physics and the Entanglement of Matter and Meaning.* Durham, NY: Duke University Press.

Barthes, R. (1984) *The Rustle of Language* (R. Howard, trans.). Berkeley: University of California Press.

Bayley, A. (2016) Trans-forming higher education. *Performance Research* 21 (6), 44–49.

Blackledge, A. and Creese, A. (2017) Translanguaging and the body. *International Journal of Multilingualism* 14 (3), 250–268.

Bradley, J. and Moore, E. (2018) Resemiotization and creative production: extending the translanguaging lens. In A. Sherris and E. Adami (eds) *Making Signs, Translanguaging Ethnographies: Exploring Urban, Rural and Educational Spaces* (pp. 91–111). Bristol: Multilingual Matters.

Bradley, J., Moore, E., Simpson, J. and Atkinson, L. (2018) Translanguaging space and creative activity: Theorising collaborative arts-based learning. *Language and Intercultural Communication* 18 (1), 54–73.

Callaghan, J., Moore, E. and Simpson, J. (2018) Co-ordinated action, communication, and creativity in basketball in superdiversity. *Language and Intercultural Communication* 18 (1), 28–53.

Clarke, M. and Hennig, B. (2013) Motivation as ethical self-formation. *Educational Philosophy and Theory* 45 (1), 77–90.

Frimberger, K. (2013) Towards a Brechtian research pedagogy for intercultural education: Cultivating intercultural spaces of experiment through drama. PhD thesis, University of Glasgow.

Frimberger, K., White, R. and Ma, L. (2017) 'If I didn't know you what would you want me to see?': Poetic mappings in neo-materialist research with young asylum seekers and refugees. *Applied Linguistics Review* 9 (2–3), 391–419.

Gotman, K. (2016) Translatio. *Performance Research* 21 (5), 17–20.

Harvey, L. (2014) Language learning motivation as ideological becoming: Dialogues with six English-language learners. PhD thesis, University of Manchester.

Harvey, L. (2015) Beyond member-checking: A dialogic approach to the research interview. *International Journal of Research and Method in Education* 38 (1), 23–38.

Harvey, L. (2016) 'I am Italian in the world': A mobile student's story of language learning and ideological becoming. *Language and Intercultural Communication* 16 (3), 368–383.

Harvey, L. (2017) Language learning motivation as ideological becoming. *System* 65, 69–77.

Harvey, L. (2018) Adapting intercultural research for performance: Enacting hospitality in interdisciplinary collaboration and public engagement. *Arts and Humanities in Higher Education* 17 (4), 371–387.

Harvey, L., McCormick, B., Vanden, K., Suarez, P. and Collins, M. (2019) Theatrical performance as a public pedagogy of solidarity for intercultural learning. *Research for All* 3 (1), 74–90.

Harvey, L., McCormick, B. and Vanden, K. (2019) Becoming at the boundaries of language: Dramatic Enquiry for intercultural learning in UK higher education. *Language and Intercultural Communication*. DOI: 10.1080/14708477.2019.1586912

Hicks, D. (2000) Self and other in Bakhtin's early philosophical essays: Prelude to a theory of prose consciousness. *Mind, Culture, and Activity* 7, 227–242.

Huang, Z.M. and Fay, R. (2016) Knowledge Transcreation. Unpublished manuscript, University of Manchester.

Iedema, R. (2003) Multimodality, resemiotization: Extending the analysis of discourse as multi-semiotic practice. *Visual Communication* 2 (1), 29–57.

Jones, A. (2016) Introduction. *Performance Research* 21 (5), 1–11.

Kubanyiova, M. and Crookes, G. (2016) Re-envisioning the roles, tasks, and contributions of language teachers in the multilingual era of language education research and practice. *The Modern Language Journal* 100 (1), 117–132.

Mylona, S. (2016) Trans-ing. *Performance Research* 21 (5), 65–67.

10 The Hyper-Local Development of Translanguaging Pedagogies

Kendall A. King and Martha Bigelow

This chapter examines the potential of translanguaging policies and pedagogies to support refugee-background, English learning adolescents and young people. We first examine the empirical and theoretical basis supporting the use of translanguaging approaches for students with interrupted or limited formal education and emerging literacy skills, and then consider four specific policies and pedagogies enacted in one US context. As evident throughout this book, notions of translanguaging have been variably interpreted and concomitantly, increasingly influential. In broad terms, translanguaging is a conceptual innovation that embraces positive, holistic, and fluid notions of bilingualism and multilingualism, and takes seriously the productive potential of using students' languages as resources for communication and full engagement in educational contexts.

Over the last decade, as the notion of translanguaging has been taken up by a growing number of scholars (e.g. Caldas & Faltis, 2017; García *et al.*, 2017), there has been debate over how the term is defined and, in particular, its implications for past and current lines of research. Some have taken translanguaging as a psycholinguistic construct, suggesting that translanguaging, by definition, demands repudiation of the notion that multilinguals hold competencies in different, discrete so-called 'named' languages. From this vantage point, multilingual individuals have internally undifferentiated, unitary language systems (Otheguy *et al.*, 2015), or at least an integrated multilingual knowledge base (Herdina & Jessner, 2002). Empirical work in code-switching (MacSwan, 2014) and much scholarship on bilingual first language acquisition (e.g. Genesee *et al.*, 1995) refutes this unitary language take, with some scholars noting that this unitary approach undermines important arguments concerning the systematic, rule-governed nature of code-switching. This argument (and supporting research) has been important in 'talking back' to inaccurate but widely circulating derogatory notions such as 'semilingual,' often used to describe the language of emerging bilinguals.

Rather than take up those debates here, we adopt here what MacSwan (2017) has termed a multilingual perspective on translanguaging. This approach emphasizes that multilingualism is psychologically 'real' (acknowledging that speakers do move between different, and often recognizable languages), but also points to the ways that translanguaging is productive for learning and engagement in classrooms. This perspective makes clear both the long-standing language ideologies that have privileged monolingual language use (and correspondingly, diminished and problematized use of additional languages), and points to the ways in which code-switching in the classroom can productively enhance both language and content learning while also contributing to culturally sustaining pedagogies (García et al., 2017; Paris, 2012).

For instance, as evident in other chapters in the present volume (e.g. Ballena et al.; Llompart-Esbert & Nussbaum), empirical evidence points to the productivity of translanguaging practices for facilitating learning in school. This is sometimes referred to as pedagogical or intentional translanguaging (to contrast with the fluid, everyday use of languages outside of school). Productive, intentional pedagogical approaches might entail reading a text in one language, and discussing it in another, or alternating the languages used for input and output systematically. For instance, researchers in New Zealand found that Māori language literacy increased when students were permitted to use their first language (in this case, English) to discuss Māori medium texts (Lowman et al., 2007). Similarly, recent work with Deaf adults in the US suggests that translanguaging served as a bridge from American Sign Language to English literacy (Hoffman et al., 2017).

An important critique of translanguaging pedagogies to date concerns their utility within minority language contexts. As Cenoz and Gorter (2017) note, in situations where the balance of power of the languages is uneven, school is often a prime (and perhaps only) site where minority languages can be protected and practiced. There is concern that translanguaging pedagogies might impact both the language status (crowding out limited 'space' for use of the language) and the language corpus (leading to loss of particular forms or structures), as suggested by Basque scholars (Zarraga, 2014; in Cenoz & Gorter, 2017).

While debate over the applicability of translanguaging pedagogies continues, limited work to date on translanguaging has attended to its utility and particular value to adolescents who are still in the process of becoming literate who are also new to formal schooling. These students, sometimes known as 'Students with Limited or Interrupted Formal Education' (SLIFE), are a significant and growing part of the school-age population worldwide. The UN reports that an unprecedented 65.6 million people have been forced from home, including 22.5 million refugees, over half of whom are under the age of 18 (UNHCR, 2017). Many youths spend all or part of their school-aged years in states with non-functioning schools, or in refugee camps with uneven access to formal education, missing opportunities to acquire literacy and academic content prior to adolescence and prior to final settlement. These students then face the challenge of acquiring the language of their new host country, learning how to read and write most often in a language they do not speak or understand, and acquiring missed academic content and skills across multiple subjects (DeCapua *et al.*, 2007). Upon arrival to their place of resettlement, they must receive specialized instruction or they will likely fall further behind.

On both theoretical and empirical grounds, there is a strong argument for policy acceptance and pedagogical integration of translanguaging approaches among refugee/immigrant and English learning adolescents. Indeed, pedagogies and policies that take up multilingual perspectives on translanguaging have the potential to foreground students' linguistic and cultural assets and to transform classroom processes. Of all student populations, SLIFE are poised to potentially benefit *the most* from translanguaging approaches. Pedagogical approaches to translanguaging allow teachers to support students as they engage with academically complex content and texts; to provide opportunities for students to develop linguistic practices needed for specific academic contexts; to create space for students' growing bilingualism and particular ways of knowing; and to facilitate students' socio-emotional development and bilingual identities (García *et al.*, 2017: ix). While important for all learners, these pedagogies are arguably even more critical for SLIFE as these students often arrive at school with extensive, rich life experiences and skills, but often limited knowledge of how to 'do school'. Taking up translanguaging approaches has the potential to help bridge these sets of experiences, and accelerate and deepen engagement with academic texts and acquisition of academic skills.

Furthermore, translanguaging approaches dovetail with the extensive work confirming the value of instruction via students' first language. For more than six decades, researchers and policymakers around the world have recognized the benefits of multilingual education. For example, in 1953 UNESCO published an important report on the *Use of Vernacular Languages in Education*, which continues to be the most frequently cited document on education. These UNESCO recommendations stress the benefits and importance of educating students through a language that they comprehend. For students who come from minority language backgrounds, this means educating them in and through their mother tongues. Sixty years on from this foundational report, a substantial body of research indicates that instruction via students' native languages enhances the development of content knowledge, skills, and literacy (e.g. Thomas & Collier, 2002).

Moreover, the maintenance and development of two or more languages over time is associated with multiple health and cognitive advantages, and positive emotional and social development. While we do not suggest that instructional use of students' native languages amounts to translanguaging pedagogies, we point out that the argument for translanguaging instructional approaches rests in part on, and is bolstered by, the significant body of work supporting native language use.

Local and Global Trends

Despite the empirical base supporting both the use of students' first languages and the productive incorporation of translanguaging pedagogies, educational institutions often treat such approaches as experimental. This skepticism has been well documented in international, comparative education work. For instance, Benson (2004), drawing on decades of non-governmental organization (NGO) experience and the cases of Bolivia and Mozambique in particular, notes the tendency of national education systems to 'get stuck' at the experimental phase, testing the effectiveness of programs at times for decades and delaying full implementation. Overall, despite significant policy, research and demographic trends to the contrary, programs that systematically draw on students' mother tongues and allow for pedagogical translanguaging remain the exception, both worldwide, and as illustrated below, in our own context of work.

The US state of Minnesota is distinguished by a long history of welcoming refugees, including large numbers of Hmong in the 1970s, and East Africans starting in the 1990s and continuing to the present. The state is presently home to the largest population of Somalis in North America (upwards of 50,000), and the state serves more than 21,500 school-age students who speak Somali at home (Minnesota Department of Education, 2016). Limited access to education pre- and during migration have resulted in a high incidence of low print literacy skills among many East African adolescents (Abdi, 2007). As a result of these demographics, the state has extensive professional expertise in working with refugee-background adolescents new to formal schooling, and local experts such as Drs Patsy Egan, Jen Vanek and Jill Watson, among others, are national and international leaders in this area.

In 2014, Minnesota passed progressive language education policy, known as the 2014 Learning English for Academic Proficiency and Success (LEAPS) Act (Minnesota Department of Education, 2018). This law revises many statutes to place greater emphasis on English learners. The law frames multilingualism as an asset for Minnesota and sets a high bar for native language use and instruction among English Learners (ELs). The legislation explicitly draws from a 'language-as-a-resource' perspective (Ruiz, 1984). For instance, the legislation suggests new procedures for assessing ELs and teacher re-licensure requirements that focus on teaching ELs; it also specifies features of programming that use community and cultural assets as a starting point.

As evident below, implementation to date has been restrained and partial (King & Bigelow, 2017a). Overall, efforts have tended to focus narrowly on compliance to the letter, rather than the spirit, of the law. For instance, the law indicates that districts should support literacy development in students' native languages, and develop assessments to that end. However, the language of the law contains many opt-outs. For instance, 'districts *are encouraged* to use strategies that teach reading and writing in the students' native language and English at the same time' (122A.06, subdivision 4). Another example: teacher 'preparation programs must include instruction on meeting the varied needs of English learners, from young children to adults, in English and, *where practicable*, in students' native language' (122A.14, subdivision 2) [emphasis ours]. In both cases, use of students' native language is not binding but 'encouraged' through the phrase '*where practicable*'. Given these opt-outs, and a local cultural context that prioritizes conflict avoidance, policy makers have narrowly interpreted the law in ways that emphasize technical compliance, but do not directly require instruction that develops students' native languages as an unequivocal mandate (King & Bigelow, 2019).

This tendency, while not unique to our context, is even more troubling in light of the substantial history of bilingual education in our region. For instance, during initial implementation of LEAPS, native language education and translanguaging approaches were frequently positioned by state and local leaders as novel, experimental, or logistically impossible. Like many states, local districts developed school-based bilingual programs that enjoyed official support and recognition over decades. In fact, Minneapolis Public Schools, the largest district in the region, had established bilingual education programs in the 1980s at particular school sites (personal communication, T. Siguenza, 14 July 2017). In 2006, this district developed and issued a policy on 'Bilingual Student Education' (Policy 6280). The stated purpose was to recognize 'students' native languages as an asset to be built upon and maintained for education success'

(Policy 6280: 1). This was to be accomplished by 'using instruction in students' first language to teach grade-level content (bilingual instruction)' (Policy 6280: 1). Funding and systematic support was never institutionalized, and while (to our knowledge) the policy has not been revoked, it has also never been invoked. As elsewhere, through the discursive and semiotic process of erasure (Irvine & Gal, 2009), the programs have disappeared from public discourse and administrative memory. As a result, multilingual policies that would support translanguaging pedagogies are constructed, yet again, as 'new'. Explanations (or excuses) for the lack of bilingual programming are all too familiar: too many different student languages, lack of licensed bilingual teachers, mobility of families, lack of curriculum, or lack of long-term administrative and public support (Halford, 1996). As we illustrate below, efforts to draw on students' native languages or to develop translanguaging approaches tend to be re-built from the ground up at a hyper-local level. We offer examples below, and follow with ideas of how to break this cycle.

Hyper-Local Development of Translanguaging Approaches

Here we highlight small-scale, hyper-local efforts to adopt translanguaging approaches. 'Hyper-local' is a term that is commonly used in marketing and journalism research. We use it here to signal our focus on a small region, and particular constellations of people, educational policies and geographies, and to point to the ways in which translanguaging pedagogies are often taken up locally. We describe four such small-scale collaborative efforts, their uptake, and their impact.

Translanguaging pedagogies for initial literacy development

This first example comes from our long-term, collaborative work with a multilingual classroom high school teacher, Abdiasis Hirsi (King *et al.*, 2017). During the 2015–2016 academic year, we visited Mr Hirsi's class approximately twice per week, observing the same 90-minute literacy block each visit. His students were beginning-level English students, all of whom were new to formal schooling, new to print, and new to English. All students had refugee backgrounds and were between the ages of 16 and 22; for most of the year, all students were native Somali speakers. Mr Hirsi often moved between Somali and English for informal conversation with students, but used English only for all instructional activities (e.g. decoding and spelling practice, sight word review). Students worked extensively with *Sam and Pat* texts (Hartel, 2005), practicing individual words and then reading entire sentences individually and collectively, sometimes many times each day. Classroom observations suggested that students were able to memorize entire sentences, but meaning remained opaque and decoding of individual letters or words was uneven at best. For instance, after weeks and perhaps one hundred (monolingual English) readings of *Sam and Pat* texts, students remained unclear on the basic meaning of them; for instance, who 'Sam' and 'Pat' were (e.g. randomly pointing to male or female drawings meant to illustrate the text) and to whom (and what) the high-frequency words (e.g. 'fat') referred. There was never any conversation around these texts as they functioned largely as exercises in decoding and without content that inspires conversation.

Mid-year, after collaboration and discussion with us, Mr Hirsi began to adopt a multilingual approach to teaching basic literacy, similar to the 'language experience' approach (Hall, 1970) and, as he told us, the way he taught Somali literacy, pre-migration. Using this approach, Mr Hirsi showed a photo to the class (often projected on screen), and students then discussed the content and context of the photo. Photos varied from everyday pictures of Somali families around the city, to camel trading in East Africa. These highly animated conversations took place mostly in Somali, with occasional use of English words (e.g. 'cookie'). Next, Mr Hirsi and students collectively wrote 9 or 10 short sentences in Somali. Students practiced reading these in Somali across a few class periods, and then these sentences were translated into English. These English texts became the focus of English reading practice, and choral reading in particular, with substantial strategic and social-emotional support, and at points, simultaneous translation in Somali.

Our discourse analysis of these events, and student-led choral reading episodes in particular (King *et al.*, 2017), illustrated how this activity allowed for ample talk and interaction around performance of this text (see Image 10.1). Students created and took up opportunities to provide and receive feedback, recasts, and metalinguistic commentaries, and were highly engaged in talk in English and Somali as audience members and participants. Furthermore, the bilingual discussions around these everyday photos were fast-paced and quick-witted. As an example, in discussing the photo of man by his car, a student noted the man looked happy and the students were asked why that might be. One noted it might be because there was no snow (a nod to Minneapolis weather). Another student responded that the man was obviously happy because he had found a good parking space. This comment, which elicited laughter from all, drew on and extended students' meaning-making skills from this particular image to our local, immediate urban context, in which finding street parking and snow are regular challenges. This approach to multilingual creation and sense-making with text is not radical or novel; it is just a simple pedagogical shift that is linguistically inclusive, efficient and effective.

Image 10.1 Photo of text and parking picture used in class

Despite the success of these efforts, in some respects, this approach was in many respects a policy failure. At the end of the academic year, the teacher was informed his contract might not be renewed due to union rules, and the initial literacy class might be regrouped. This is an outcome that is reminiscent of Hornberger's (1987: 205) report of 'pedagogical success, but policy failure' in Peru, which documented similar successes in using Quechua to teach literacy, but challenges in maintaining these practices. In the end, Mr Hirsi was able to keep his position, but program instability brought challenges to maintaining, much less expanding, this hyper-local translanguaging effort.

Translanguaging pedagogies for critical media skills

Our second pedagogical example draws from our collaborative work with other educators who serve a similar population of adolescent SLIFE (Bigelow *et al.*, 2017). Students were intermediate-level English learners of refugee background, but many had been in the US for several years and acquired basic print literacy. We examined how a summer school critical literacy curriculum unit, which used Facebook as a tool for interaction and publishing student work, served as a context for the Somali and Spanish speakers to interact in their native languages and participate in English literacy development. The curriculum was designed to engage youth not only in multilingual language interaction, but also to inspire youth to produce and discuss texts critically. We introduced Facebook as a tool by first creating a 'secret group' for the whole class. This secret group was only visible to classroom participants and not to other Facebook friends or networks. We then opened additional secret groups for peers of the same language background to use the space to support their writing process. The culminating project was a digital text that represented some aspect of the students' cultures and experiences. We wanted students to engage in critical analyses of texts (including visual images) and contribute to the body of multimodal artifacts online, including memes, vlogs, images, and current social media conversations. We designed instruction in ways that encouraged youth to use all their cultural, inter-personal, digital and linguistic resources throughout the process and in final products – which were highly visual, included varying quantities of text in English and were accompanied by an oral presentation to the class.

Our analysis revealed the ways in which students moved back and forth between languages (and media) for a range of purposes, even in the 'secret' language-specific Facebook groups. As an example, in the following excerpt from a post, the participant began with a greeting in Somali (*Wallalaha*), switched to English (*Kennedy international high school welcome to high like me if like. or help us comments OK please*) and then switched back to Somali (*fadlan kqsooqayb gala schoolkan ooy waxbarashadiisu aad iyo aad usarayso*), which roughly translates to 'please, participate in this school, which has high academic standards.'

> Wallalaha Kennedy international high school welcome to high like me if like. or help us comments OK please fadlan kqsooqayb gala schoolkan ooy waxbarashadiisu aad iyo aad usarayso. (11 July 2015, Sahara)

This mix of Somali and English functioned to encourage peers to participate both in 'liking' the post and participating in the school. The post affirmed the quality of the high school, which might serve to legitimize both activities.

More broadly, findings pointed to how these students used all of their languages in both inter- and intra-culturally sophisticated ways. Their multilingual Facebook posts, we speculate, have many possible motivations and explanations. First, many multilingual youth are accustomed to communicating with other multilingual youth, who leverage all of their linguistic skills for a wide range of communicative purposes, including humor, emphasis, and evoking religion. This is evident in the multiple ways youth used language in the Somali secret group. For instance, the names they chose often carried deeper meaning in Somali. One male student's chosen name signals 'flying', which speaks to experiences of travel. The use of poetry and specifically metaphor in Somali language is common and often shows one's mastery of the language (Andrzejewski, 1972). Youth are also accustomed to being linguistically inclusive, which seemed to inform some of their translanguaging practices. In other words, the use of English in some of their Facebook posts, even in their same-language secret groups, may have been an effort to not exclude those who had said they did not know how to read their common language (e.g. Somali).

Overall, use of the native languages provided opportunities for engagement with course curriculum and activities. As an example, our initial work (in English) resulted in Facebook posts that were largely superficial representations of culture: food, fashion, houses and henna. With deeper engagement through use of students' native languages, and in particular an emotional discussion of a 'Somali house' located in a refugee camp (deeply upsetting to students that this was entitled 'Somali house'), students were able to move on to more complex and nuanced discussions of culture in the second part of the course.

This project demonstrated the synergy generated by encouraging youth to employ remixing and translanguaging together in support of their learning, especially when using social media as an instructional tool/venue. Social media, by definition, is participatory, and requires viewing literacy as a social practice and opportunities to express a 'critical reading of reality' (Freire & Macedo, 1987: 36). Affording learners the opportunities to draw on all literacy and linguistic resources in support of their participation can help them develop an awareness of their ability to 'contribute to collective intelligence' (Pegrum, 2011: 9). Through this project, we saw the potential for pedagogies involving social media to foster the sort of engagement that permits refugee and immigrant youth to use all of their linguistic and cultural resources to create content not only for their peers locally, in class, but for multiple global audiences. This potential was possible through an experimental summer school class; however, to our knowledge (and despite efforts to share curriculum and lead development work), this sort of curriculum has not been adopted more broadly in state schools.

Native language literacy assessments

We now turn to two policy initiatives. The first of these grew out of our collaboration with the Minneapolis public schools and in particular, consulting work at their intake center, where all new students come to register, be assessed, and then placed in a school and often, an ESL level within that school. Our work revealed that the English language assessment used in the state (and widely across the country), known then as the W-APT, or the 'the screener,' was a suboptimal assessment for many students (King & Bigelow, 2017b). The test allowed the district to comply with federal ESL screening mandates and provided administrators with a rough sense of students' English language proficiency across the domains of reading, writing, speaking and listening. However, the test failed to effectively differentiate between those with previous schooling outside the US and in a language other than English, and those with no schooling whatsoever. Students with remarkably varied literacy skills (in languages other English) earned nearly identical scores. The test tended to underestimate students' skills and be insensitive in particular to students who are preliterate or acquiring literacy. As a result, students are often initially placed in classes not appropriate to their skill level, thus slowing down their achievement and learning.

Current research in meeting the needs of English learners rests on a translanguaging framework which 'ensures that the students' different home language practices are not only validated, but also used and leveraged for academic purposes' (García & Hesson, 2015: 221), including critical and creative thinking. An important first step in this direction is the development and implementation of assessments that reflect students' multilingual and multiliterate capacities. To that end, in collaboration with students, teachers and administrators across the state, we developed the Native Literacy Learning Assessment (NLLA). This test, which is free and available state-wide, provides administrators and teachers with information about students' reading and writing skills in their first language. It is currently available in Spanish, Somali, Oromo, Arabic, Chinese, Swahili, English and Amharic, and does not require the test administrator to be knowledgeable of testing language.

While the NLLA is essentially a monolingual test, it allows students to demonstrate formal literacy skills in any language. The scoring system differentiates across students with no, partial, and extensive formal education, and scores literacy skills based largely on writing fluency and quantity of text production. In this (limited and imperfect) sense, it takes a translanguaging approach in that students are evaluated on their demonstration of competencies in languages other than English, and demonstrate those competencies in whichever languages they have the most print literacy skills. This is a simple innovation, but has the potential to complement other placement tools that are only in English. Teachers would need to follow up with other native language assessment tools in order to obtain a more complete and fine-grained assessment of the multiple native language literacies a new student might possess, but the NLLA offers a small step in the right direction.

Multilingual seals and certificates

The final initiative we discuss here is part of the LEAPS legislation outlined above; as part of this new law, the state established bilingual and multilingual recognition seals. State-issued seals and certificates of biliteracy are increasingly common in the US: California first passed legislation supporting biliteracy seals in 2012. Since then, 25 states have followed and at least nine others are in the early stages of implementation. While seal legislation is often presented as an important advance in promoting foreign language teaching (for native-English speakers) and recognizing the language skills of heritage speakers (for emergent bilinguals), as we suggest below, uptake to date has been uneven and presents challenges in terms of developing assessments that are appropriate, valid and reliable across populations and languages.

This Minnesota legislation grants students a gold or platinum seal or certificate if they can demonstrate proficiency in a language other than English. Within Minnesota, we (Schwedhelm & King, in press) examined how the bilingual and multilingual seals have been implemented thus far; and in particular, who has benefited from this legislation to date. Data were collected through close discourse analysis of the state law, quantitative analysis of district and state data, and interviews (10 in total) with key policy actors, teachers and students in 2017.

Findings suggest that while the legislation provides important ideological and implementation space for multilingualism, uptake has been restricted by administrative and technical challenges with uneven access to the seals and certificates across rural, urban and suburban schools and languages. While the seals and certificates are issued by the Minnesota Department of Education, administration of the exams, reporting of the scores, and preparation for the assessments is all handled at the district, and in some cases, at the school and classroom level. In practice, this means that whether a student has access to the seal largely depends on district-level programming. In smaller districts in particular, access was dependent upon a foreign language teacher or director who was knowledgeable about (and interested in) promoting the seal, preparing students for the exam, and setting up proficiency tests. In this sense, uptake was hyper-local.

Early implementation years were also marked by administrative challenges, including how to add a state 'seal' to varied types of online high school transcripts. Furthermore, and more importantly, no state funding was set aside to pay for the exam, with these costs carried by the districts. In some smaller districts, budgets for English learners were used to cover the cost of the tests. These limited resources required districts to make difficult decisions about who to serve with these funds and whether to limit access.

An additional challenge has been developing assessments in languages that are less widely used and taught (at least in the US), such as Oromo, Tamil and Karen. The Minnesota Department of Education has focused on developing tests and training raters for the 12 most widely spoken languages in the state. While Spanish and other common world language assessments were available from the first year of implementation (2015), this was not the case for many Minnesota immigrant and Indigenous languages. However, in recent years the number of seals and certificates awarded to other languages has grown. In Spring of 2017, seals and certificates were awarded for newly developed assessments in Hmong (21), Karen (36), Somali (17) and Tamil (2). While this represents progress, most seals were awarded to students of traditionally taught world languages, including Chinese (62), French (208), German (15), Japanese (22) and Spanish (847). This imbalance is particularly notable given the large number of Somali, Hmong and Karen students in the state, with thousands of students using these languages at home (US Census, 2017).

Lastly, the seals and certificates have brought issues of translanguaging to the forefront. First, a growing number of students, teachers and advocates have questioned the monolingual norm inherent in the proficiency assessments. The ACTFL guidelines are not reflective of what it means to be communicatively competent in many local communities within Minneapolis and beyond. Second, the state legislation requires students earning the seal or certificate to demonstrate not only world language proficiency, but also English language proficiency. As one administrator in a large multilingual district noted, 'this is an equity issue because only the EL kids have to be taking ACCESS [a standardized test] to demonstrate English Language Proficiency', not non-EL students. Both issues point to the monolingual norms and inequities across student groups underlying the legislation. More broadly, our analysis suggests that implementation has been hyper-local with uptake and manner of implementation dependent on local teachers and districts. This work points to the need for leadership and public advocacy for the seals to recognize their full potential.

Conclusion: Normalizing Translanguaging Practices Locally

Translanguaging pedagogies and policies at school, while theoretically or equity informed, are enacted by individuals, within a complex web of hyper-locally determined practices, within a deep history of inequality. Therefore, when educators and policymakers strive to enact educational reforms in the best interest of multilingual, refugee-background youth, we must be vigilant of ways that new dynamics of inequality can be created. For example, do our new Seals of Biliteracy privilege those who have had formal (literacy) instruction in their non-English language, or does this legislation drive opportunities for students to acquire the required levels of biliteracy? The answer to this question is yet to be seen, and dependent upon, for instance, how many innovations and investments are made to train raters across the different languages, and how many native English speakers receive the Seals relative to (former or current) English learners. We hope that the seal legislation, in the spirit of 'multilingualism for all', does not fall prey to the forces of globalization which favor powerful languages, spoken by students from the powerful side of race, ethnicity and class divides, from schools with the most robust support for world language education.

More broadly, we hope that local decisions about how to promote biliteracy in 'smaller' languages, immigrant languages, and languages learned entirely through the oral modes will be supported in concrete and sustainable ways that do not require continual (re)invention. Hiring teachers, buying materials, training raters, and enlisting community members in the process are actions which take effort, potentially confront resistance, and certainly call for seeing the development of biliteracy in a home or ancestral language as being as important as developing English literacy and integral to the public schooling experience. The fact that the two are very closely intertwined is often lost in locally informed decision making about the sort of schooling immigrant youth need.

We also recognize that schools are increasingly asked to do more with less funding. There are constant accountability pressures across multiple student groups, including English learners (or multilinguals). As Alexander and Jang (2017) point out, policymakers typically strive to base funding decisions on the fair distribution and efficient use of public resources, and we note that this may involve uneven distribution of resources to reach educational objectives (Alexander, 2012). For example, while total expenditures per Minnesota student increased from 2003–2011, levels of inequalities also increased (Alexander & Jang, 2017). Disappointingly, districts with more English learners had lower expenditures per pupil, even though they are mostly in districts reporting high rates of poverty and in markets with high labor costs (Alexander & Jang, 2017). This double bind does not seem to have an easy answer, but we suggest that educators, in collaboration with immigrant communities and school leadership, might be able to leverage research on the academic benefits of multilingualism together with Minnesota's new legislation that explicitly mandates instruction that includes students' languages other than English (King, 2018). The LEAPS Act could support the institutionalizing of multilingual practices in Minnesota, statewide, without each school or district needing to create their own ad-hoc systems locally or hyperlocally.

In our work in schools, we often wonder if the root causes of the barriers to linguistically inclusive teaching involve 'the skill or the will' among us to implement what we know works for language minority SLIFE. In other words, do we need to broaden our *skill* with multilingual teaching strategies and program design? Or do we need to develop the *will* to dismantle English-only approaches to immigrant/refugee education? While we recognize that there are many excuses for not implementing multilingual programming, the argument that such shifts in programs for English learners are empirically risky is unfounded. We have ample evidence that it is not only advantageous to teach in ways that are linguistically inclusive; furthermore, we argue that it is the right of our students to be educated this way, now recognized and required in state legislations (LEAPS). Standing on strong legal ground (e.g. Lau v Nichols), these pedagogies are legalized, legitimized and mandated in the state of Minnesota (Sugarman & Widess, 1974). That said, we have shown how laws about language policy can be subverted through the very text of the law, through implementation challenges, and the degree to which the local culture allows for a definitive enactment of the spirit of the law. We need our Minnesota Department of Education, as the main driver in implementation, to be bold in supporting and demanding hyper-local implementation efforts.

In the absence of uniform statewide changes to language instruction, we have documented the ease with which translanguaging pedagogies and multilingual assessment can be implemented at the local level. We advocate for more spaces where pre-service and in-service educators and administrators can imagine multilingual teaching in ways that use their existing skills/resources and work against the pervasive belief of the impossibility of multilingual teaching.

References

Abdi, C. M. (2007) Convergence of civil war and the religious right. *Signs: Journal of Women in Culture and Society* 33 (1), 183–207.

Alexander, N.A. (2012) *Policy Analysis for Educational Leaders: A Step-by-Step Approach*. New York, NY: Pearson.

Alexander, N.A. and Jang, S.T. (2017) Equity and efficiency of Minnesota educational expenditures with a focus on English learners, 2003–2011: A retrospective look in a time of accountability. *Education Policy Analysis Archives* 25 (16). http://dx.doi .org/10.14507/epaa.25.2811

Andrzejewski, B.W. (1972) Poetry in Somali society. In J. B. Pride and J. Holmes (eds) *Sociolinguistics: Selected Readings* (pp. 5–8). Harmondsworth: Penguin.

Benson, C. (2004) Bilingual schooling in Mozambique and Bolivia: From experimentation to implementation. *Language Policy* 3 (1), 47–66.

Bigelow, M., Vanek, J., King, K.A. and Abdi, N. (2017) Literacy as social (media) practice: Refugee youth and native language literacy at school. *International Journal of Intercultural Relations* 60 (Sept), 183–197.

Caldas, B. and Faltis, C. (2017) Más allá de poly, multi, trans, pluri, bi: ¿De qué hablamos cuando hablamos de translinguismo. *NABE Journal of Research and Practice* 8 (1), 155–156.

Cenoz, J. and Gorter, D. (2017) Minority languages and sustainable translanguaging: Threat or opportunity? *Journal of Multilingual and Multicultural Development* 38 (10), 901–912.

DeCapua, A. Smathers, W. and Tang, L.F. (2007) Schooling, interrupted. *Educational Leadership* 64 (6), 40–46.

Freire, P. and Macedo, D. (1987) *Literacy: Reading the Word and the World*. South Hadley, MA: Bergin and Garvey Publishers.

García, O. and Hesson, S. (2015) Translanguaging frameworks for teachers: Macro and micro perspectives. In A. Yiakoumetti (ed.) *Multilingualism and Language in Education* (pp. 221–242). Cambridge: Cambridge University Press.

García, O., Johnson, S.I. and Seltzer, K. (2017) *The Translanguaging Classroom: Leveraging Student Bilingualism for Learning*. Philadelphia, PA: Caslon, Inc.

Genesee, F., Nicoladis, E. and Paradis, J. (1995) Language differentiation in early bilingual development. *Journal of Child Language* 22 (3), 611–631.

Halford, J.M. (1996) Bilingual education: Focusing policy on student achievement. *ASCD* 4 (March). www.ascd.org/publications/newsletters/policy-priorities/mar96/num04/Focusing-Policy-on-Student-Achievement.aspx (accessed January 2020)

Hall, M.A. (1970) *Teaching Reading as a Language Experience*. Columbus, OH: Charles Merrill.

Hartel, J.A. (2005) *Sam and Pat*. Boston, MA: Heinle ELT.

Herdina, P., and Jessner, U. (2002) *A Dynamic Model of Multilingualism: Perspectives of Change in Psycholinguistics*. Clevedon: Multilingual Matters.

Hoffman, D., Wolsey, J.-L., Andrews, J. and Clark, D. (2017) Translanguaging supports reading with deaf adult bilinguals: A qualitative approach. *The Qualitative Report* 22 (7), 1925–1944.

Hornberger, N.H. (1987) Bilingual education success, but policy failure. *Language in Society* 16 (2), 205–226.

Irvine, J. and Gal, S. (2009) Language ideology and linguistic differentiation. In A. Duranti (ed.) *Linguistic Anthropology: A Reader* (pp. 402–434). NY: Wiley.

King, K. (2018) Embrace multilingualism as a goal for all Minnesota students. *Star Tribune* (Opinion/Exchange). www.startribune.com/embrace-multilingualism-as-a-goal-for-all-minnesota-students/502118462/ (accessed January 2020).

King, K., Bigelow, M. and Hirsi, A. (2017) New to school and new to print: Everyday peer interaction among adolescent high school newcomers. *International Multilingual Research Journal* 11 (3), 137–151.

King, K.A. and Bigelow, M. (2019) The politics of language education policy implementation: Minnesota (not so) nice? In T. Ricento (ed.) *Language and Politics in the U.S. and Canada* (pp. 192–211). Cambridge: Cambridge University Press.

King, K.A. and Bigelow, M. (2017a) Minnesota (not so) nice?: LEAPS policy development and implementation (Invited article). *MinneTESOL Journal*. http://minnesoljournal.org/fall-2017-issue/minnesota-not-nice-politics-language-education-policy-development-implementation (accessed January 2020)

King, K.A. and Bigelow, M. (2017b) The language policy of placement tests for newcomer English learners. *Educational Policy*. http://doi.org/10.1177/0895904816681527

Lowman, C., Fitzgerald, T., Rapira, P. and Clark, R. (2007) First language literacy skill transfer in a second language learning environment. *SET* 2, 24–28.

MacSwan, J. (2014) *Grammatical Theory and Bilingual Codeswitching*. Cambridge, MA: MIT Press.

MacSwan, J. (2017) A multilingual perspective on translanguaging. *American Educational Research Journal* 54 (1), 167–201.

Minneapolis Public Schools (June 13, 2006) Bilingual Student Education. School Board Policy 6280.

Minnesota Department of Education (2016) *Data Reports and Analytics*. https://public.education.mn.gov/MDEAnalytics/DataTopic.jsp?TOPICID=2 (accessed January 2020)

Minnesota Department of Education (2018) *LEAPS Act*. https://education.mn.gov/MDE/dse/el/leap/ (accessed January 2020)

Otheguy, R., García, O. and Reid, W. (2015) Clarifying translanguaging and deconstucting named languages: A perspective from linguistics. *Applied Linguistics Review* 6 (3), 281–307.

Paris, D. (2012) Culturally sustaining pedagogy: A needed change in stance, terminology, and practice. *Educational Researcher* 41 (3), 93–97.

Pegrum, M. (2011) Modified, multiplied, and (re-) mixed; Social media and digital literacies. In M. Thomas (ed.) *Digital Education: Opportunities for Social Collaboration* (pp. 9–35). New York: Palgrave Macmillan.

Ruiz, R. (1984) Orientations in language planning. *NABE Journal* 8 (2), 15–34.

Schwedhelm, M. and King, K.A. (in press) The neoliberal logic of state seals of biliteracy. *Foreign Language Annals*.

Sugarman, S.D. and Widess, E. (1974) Equal protection for non-English-speaking school children: Lau v. Nichols. *California Law Review* 62, 156–182.

Thomas, W. and Collier, V. (2002) *A National Study of School Effectiveness for Language Minority Students' Long-term Academic Achievement*. Berkeley, CA: CREDE

UNESCO (1953) *The Use of Vernacular Languages in Education*. Paris: UNESCO.

UNHCR (2017) *Figures at a Glance*. www.unhcr.org/en-us/figures-at-a-glance.html (accessed January 2020)

US Census (2017) *Minnesota Lg Data*. https://factfinder.census.gov/faces/tableservices/jsf/pages/productview.xhtml?src=bkmk (accessed January 2020)

US Department of State (2016) *Protracted Refugee Situations*. www.state.gov/j/prm/policyissues/issues/protracted

Zarraga, A. (2014) Hizkuntza-ukipenaren ondorioak. www.erabili.eus/zer_berri/muinetik/1410777403/1410864140 (accessed January 2020)

11 Collaborative and Participatory Research for Plurilingual Language Learning

Júlia Llompart-Esbert and Luci Nussbaum

Introduction

In Barcelona, as in many cosmopolitan cities, the school population is highly linguistically and culturally diverse, especially in certain neighbourhoods. Schools teach Catalan and Spanish (official languages in Catalonia) and one or two European languages, mainly English. At some schools, pupils' heritage languages are also taught, but, until the new provisions of 2019, outside the scheduled class time. Although the educational authorities consider, as reflected in official documents, that the linguistic heritage of students is an asset to be valued, teachers generally proscribe the use of diverse languages in the classroom.

The ethnographic observations carried out in various educational institutions by our research team (Corona *et al.*, 2013; Llompart-Esbert, 2017; Nussbaum & Unamuno, 2006) show that in public and private conversations, students are highly competent at using their linguistic and communicative repertoires for various practical purposes, including acquiring new linguistic resources.

In this chapter, we discuss the process of an educational project carried out by the first author of the chapter, a teacher and high school pupils in which students investigate and describe their own communicative practices in various social domains. The environment and the characteristics of the educational institution where the research took place are then presented. Following this, we justify the didactic choices that guided the educational project in the light of certain conceptualisations of plurilingualism and translanguaging as social practices for language learning. Likewise, the proposed collaborative and participatory approach is discussed. The data collected during the project's implementation process, the outcomes and the way in which the actors evaluated these outcomes are then presented.

The High School Setting

The high school that is the focus of our study is located in one of the neighbourhoods of Barcelona with the highest percentages of immigrant origin population: around 43% (Idescat, 2014). Some of the students were born in Catalonia or arrived before adolescence; others arrived as teenagers. Therefore, both their schooling levels and their language skills in the school languages are diverse. The ethnographic work carried out in the high school in five classes has allowed about 35 different languages and varieties to be identified, reflecting pupils' extremely diverse linguistic trajectories. Consequently, the high school brings together students with complex and enormously rich linguistic repertoires that can include languages and varieties of their family's home countries, of previous schooling, of schooling in Catalonia, languages and varieties they use on the Internet and in video games, in religious spaces, etc.

Teachers' efforts are focused on ensuring that, at the end of their compulsory education, all students will have achieved equivalent competences in official languages, as well as some competences in the European languages that are taught at school. Since a large number of students do not acquire Catalan in the family environment, and to counteract the minority use of this language in the social surroundings, the educational institution is committed to teaching Catalan in all curricular areas. However, when putting this into practice, it is common that the use of other languages is proscribed, following a 'one language at a time' logic. This monolingual ideology also includes a vision of the students' plurilingual repertoires as a problem rather than as a starting point for learning (Nussbaum & Masats, 2012).

However, during our fieldwork, we observed that this monolingual rule was not applied in the so-called reception classroom, where newcomers learn Catalan and school standards. In this class, the teacher promoted plurilingual practices as a means for developing competences in the target language and as a way to empower students by giving value to their previously acquired communicative skills. This led us to question whether this procedure could be applied in a regular classroom during the teaching of curricular content and in collaboration with the teacher in charge of the subject. This was possible in a class led by a Spanish teacher.

In the following section we justify the procedure that led to the development of an educational project in which the students were the main protagonists.

Plurilingual Practices for Academic Discourse Learning

The principles and guidelines developed in Europe concerning multilingualism (the presence of two or more languages in a given society) and plurilingual competence (ability to choose and use different linguistic resources for practical purposes) have been the subject of several objections. One of these critiques focuses on the supposed static nature of plurilingual competence, and the use of indicators for its evaluation based on monolingual conceptions (Pekarek-Doehler, 2011). However, European orientations have allowed a critical reflection on school language education and profound changes in this field, especially due to the introduction, beyond grammatical knowledge, of other dimensions of language skills (discursive, sociolinguistic, pragmatic and reflexive). On the other hand, complementing the *Common European Framework of Reference for Languages* (CEFR; Council of Europe, 2001), the *Framework of Reference for Pluralistic Approaches* (FREPA) has been developed (Candelier, 2007), seeking to be an instrument to take linguistic and cultural diversity seriously in children's and adolescents' education (Candelier, 2007; Moore & Nussbaum, 2016). Published more recently, the extended version of the CEFR (Council of Europe, 2018) includes indicators of learners' capacities for exploiting their plurilingual repertoires, admitting that hybrid forms of language use (or translanguaging) are part of discursive regimes with which 21st century learners should be familiar (García & Sylvan, 2011), as is frequent in contexts of migration and mobility.

In the work we present here, some of the dimensions of these European proposals have been retained, including: (i) the recognition of plurilingual competence, understood as creative and original, in constant change, and as a basis for the development from students' informal daily communicative practices towards the academic practices demanded by educational institutions (Nussbaum, 2013); (ii) the development of knowledge and awareness regarding linguistic diversity, to give value to the sociocultural and plurilingual capital of school populations and to empower their usual linguistic practices (Moore & Nussbaum, 2016).

In this section, in the first place, we present our conceptions of plurilingual practices, on which the work we carried out with the students in the classroom is based. In the second place, some dimensions of our collaborative work with teachers are explained. In the third place, project-based work, in which students are protagonists, is presented as a framework for the development of school communicative competences.

Plurilingual practices for language learning

Sociocultural and interactionist explanations for the processes of acquiring new linguistic resources erase the borders between learning in formal contexts and language use in every day communication. Plurilingual competence is acquired through life experience, by participating in communicative practices. In fact, socio-interactionist approaches (Hall, 1993; Hellermann, 2007; Lave & Wenger, 1991; Young, 2007) understand learning as a process unfolding in interaction, rather than located in the individual's brain. From this perspective, learning is an activity that is socially situated (Mondada & Pekarek, 2004), deployed in communities of practice (Lave & Wenger, 1991), and distributed among participants.

These approaches consider *changes in ways of participating* in socially situated activities as evidence of learning. The interest in research within this tradition is not tracking changes in learners' knowledge of linguistic forms, but in the use of *all language* in interactional practices (García, 2009; Hall *et al.*, 2011; Young, 2007). Likewise, failure to learn may not be attributed to individual factors (aptitude, motivation, etc.), but to difficulties in participating in events in the classroom or outside of it. Didactic approaches must therefore favour participation promoting agency in students. The concept of translanguaging takes a similar point of view because of its potential for empowering students.

Our approach (see Llompart-Esbert & Nussbaum, 2018) adopts also the above-mentioned socio-interactionist perspectives which pay special attention to sequential occurrence of plurilingual activities and the ways, in talk-in-interaction, that learners deal both with the activity in progress and with their available repertoires to solve communicative challenges, as in the following excerpt, during which students are preparing to do fieldwork as part of the project (see the section *Documenting language uses through fieldwork*). In the excerpt, the three participants, who share Urdu, Punjabi, English, Spanish and Catalan, are deciding which materials they will use for their data collection phase.

Excerpt 11.1[1] Participants: Usama (USA), Fahad (FAH), Abdel (ABD)

01 FAH   cámara para grabar (2.7) y esto para qué necesitamos la [gra°badora°/
camera to record (2.7) and this what do we need the [re°corder° for/

02 USA   [para- para:

03   recordar
[to- to:

record

04 FAH   grabar eh: para grabar sonido/
         record eh: to record sound/
05 ABD   con la cá[mara
         with the ca[mera
06 FAH   [cám- cámara ya graba sonido
                   [cam- camera already records sound
07   (0.6)
08 ABD   pregúntalo (.) si hay no sí hay- hay cámaras que no hay graba[ción
         ask if it does (.) if there is yes the- there are cameras that don't
         re[cord sound
09 USA   [necesi-
10   esto necesitaremos para grabar
         [we ne-
         this we will need to record
((transcription of some lines ommitted))
11 ABD   °no se oye bien°
         °it is not hearable°
12   (0.3)
13 USA   OYE te está grabando
         HEY this is recording you
14   (1.5)
15 ABD   déjalo déjalo déjalo
         let it go let it go let it go
16   (0.5)
17 USA   oye no X[X
         hey don't X[X
18 FAH   [cámara para grabar
                   [camera to record
19 USA   he dicho bien no/
         i said it correctly right/
20   (0.5)
21 USA   grabar recordar no/ (0.5) ricardi' (1.1) eh (0.5) gra[bar es igual-
         to record to record right/ (0.5) ricardi' (1.1) hey (0.5)
         re[cording is the same
22 ABD   [x también
           [x also
23   necesitamos el trí-   [trí- trípo]de
         we need the tri- [tri- tripo]de
24 USA   [tío   ] (0.3) dímelo
         [man] (0.3) tell me
25 ABD   el trípode
         the tripode
26   (0.3)
27 USA   grabar es: igual: que recording (0.2) en inglés/
         to record is: like: recording (0.2) in english/
28   (0.5)
29 ABD   record yes\
30   (0.2)
31 USA   reco::rd

At the start of the excerpt, USA (lines 2–3) says why the recording device is needed, but he uses a mixed word: an English lexeme (record) and a Spanish morpheme (-ar). FAH (line 4) hetero-repairs USA by proposing to him the word in Spanish, although USA does not signal the repair. USA retains the word since, later on (lines 9–10), he uses it with the same format. In line 13, USA formulates a more complex flexion of the verb 'grabar': 'te está grabando' ('it is recording you'). In the following turns, USA looks for confirmation of the use of the verb 'to record', using several procedures: he makes a direct request (lines 19, 24, 27); he repeats the conflicting words (line 21); and he translates the word into Urdu (line 21: 'ricardi') and into English (line 27). Finally, ABD confirms the translation (line 29). Thus, it can be seen that communicative and learning activities are interwoven in interaction.

Studies on the relationship between plurilingual practices and learning have emphasised that people developing communicative expertise in a second or subsequent language systematically draw on their whole repertoire to fulfil tasks (Duff & Kobayashi, 2010; García, 2009; Li, 2011; Llompart-Esbert & Nussbaum, 2018; Lüdi & Py, 1986/2003; Lüdi & Py, 2009; Swain & Lapkin, 2000; Masats *et al.*, 2007). Such research supports the assertion that *plurilingual practices scaffold participation* in language learning, in the sense that in the first stages of learning a particular language, students might not be able to participate in the activity in progress without the support of the repertoire that they already possess.

Our collaborative research activities with teachers in educational settings (Llompart-Esbert, 2016; Nussbaum, 2017; Nussbaum & Rocha, 2008) promote didactic sequences organised around project-work in which purilingual practices are supported as a *medium* (Gafaranga & Torras, 2001) of communication in the classroom. From the point of view of the development of academic abilities (oral, written or multimodal discursive genres), plurilingual practices allow a transition between conversational activities with the teacher or classmates to more formal practices in a single language. Following Duverger (2007) and Masats and Noguerol (2016) our approach seeks 'to didactise' plurilingualism, promoting plurilingual practices from the programming of didactic sequences to classroom practices, passing through activities in which different languages are used for reading scientific texts in one given language, and writing in another language for constructing a final product.

In the research we present in this chapter, the implementation of these approaches was carried out through cooperation with a high school teacher and considering students as researchers, aspects which we expand on below.

Collaborative research

Our research aims: (i) to explore language socialisation in compulsory educational institutions; (ii) to document language learning practices; and (iii) to establish, implement, evaluate and modify holistic teaching approaches integrating languages and other subject disciplines, through projects that enhance students' role as agents/researchers and promote diverse social forms of learning (in and outside the classroom, and using technologies). These aims were pursued by adopting the principles of ethnographic research – which allowed us to obtain a dense description of everyday interactional events in educational settings through participant observation, fieldnotes, audio and video recordings, action-research – that seeks to reflect on practices in order to introduce relevant and effective modifications, anchored in specific realities and students' needs, and collaborative research.

While ethnographic research is usually conducted by researchers, and action-research by teachers, collaborative research is conducted by mixed teams that have negotiated spaces of confluence that allow access to practices of teaching/learning and their difficulties (Burns, 1999; Nussbaum, 2017; Rodriguez & Brown, 2009). In doing so, the usual borders between teachers and researchers are dissolved. The collaborating team can, during the processes and as a result of the data collection, choose *boundary objects* (Star & Griesemer, 1989; Wenger, 1998; Moore *et al.*, 2015; see also Moore & Tavares Manuel in this volume) – that is, artefacts (such as recordings and learners' productions) that allow taking a common look at them. This joint analysis promotes the creation of didactic knowledge useful for improving school practices, empowering teachers by increasing their critical capacities for theorising, and also for teacher training.

Project-work: Students as researchers

Project-based learning, which aligns with the active and critical pedagogies of Dewey, Kilpatrick and Freinet, is considered a useful tool for meaningful learning. According to Perrenoud (1999), project-work has the following characteristics: (i) it is done in a collective and collaborative way; (ii) a final product is built through the accomplishment of various tasks; (iii) this final product is real and transferable to other people, so that students understand the connection between school knowledge and social reality; (iv) it promotes knowledge about team management; and (v) it supports knowledge from different curricular areas, in addition to the abilities to self-evaluate, to cooperate and to work autonomously, which also help students raise their self-esteem.

Empowering autonomy, self-reflection and critical thinking are also some of the goals of project-based learning. Moving from the figure of the teacher as the sole bearer of knowledge to the possibility of students' own access to information modifies the roles of both learners and teachers radically. Helping students become agents of their learning can guide them to build their social world and transform it (Stetsenko, 2014).

Approaches that place students as researchers consider them active agents and see the role of education as being liberating – in the Freirean sense – and transformative. Participatory action research focuses on students' real problems, needs, desires and experiences, emphasising their active contribution and their expert roles. Reflexivity is also strengthened because students become aware of their own competences and those of others, and empowerment is encouraged because giving voice to youth allows them to resituate themselves in society, as well as in the school environment, in a new relationship with teachers and academic knowledge (Egan-Robertson & Bloome, 1998).

In our work, we are also inspired by research that situates students as ethnographers and sociolinguists (Egan-Robertson & Bloome, 1998; Lambert, 2012; Patiño & Unamuno, 2017). These studies emphasise that students' research engagement improves their self-concept, their plurilingual repertoires become visible, sociolinguistic research competence is developed, and the written language of research is learned. Taking these orientations, our proposal was intended to engage high school learners in a position of 'doing'.

Project Implementation Process and Outcomes

This chapter focuses on a project carried out with a group of 20 students from different backgrounds – which included such countries of origin as Pakistan, Bangladesh, India, Morocco, Philippines, Chile and Spain – their teacher and the researcher. The students selected the objective of discovering and documenting their language practices in different social environments, as well as investigating the activity of *language brokering* (Tse, 1996) carried out with relatives or other adults, with the final goal of making a digital poster aimed at two types of public: their own class members and a group of university Primary Education students taking a subject on school multilingualism.

In the first part of this section, we describe: (i) how languages uses were documented through fieldwork; (ii) the product of the project; (iii) the presentation of the results that young people did; (iv) the evaluation done by both the teacher and the students.

Documenting language uses through fieldwork

As students had to document their language practices, they were guided by the participating adults to conduct sociolinguistic work in order to produce a digital poster and give an oral presentation. The first session of the project was devoted to students reflecting on their communicative practices. As a result of the discussion, different settings emerged as relevant ones: the neighbourhood, the high school, the family, virtual spaces and various sites in which they carried out language brokering activities. Thus, workgroups were created – according to students' preferences or expertise – each specialising in one of these areas.

After reflecting on social multilingualism and plurilingual repertoires, students were guided through various tasks designed for collecting and processing data, as is usual in sociolinguistic research methodology:

(1) Observation about language uses in different fields, guided by questions such as: *What languages or varieties do you use? Where, when and with whom?*
(2) First reflections and planning for the collection of evidence: *What materials are needed? What evidence could be obtained?*
(3) Data collection.
(4) Group reflection and theorisation about the collected data.
(5) Preparing the final product (the digital poster and oral presentation).

Between phases 2 and 3, the students observed different posters in order to capture the peculiarities of the product that was demanded of them.

To give an example of this process, two girls and a boy of Moroccan origin, together with a girl of Pakistani origin, made up the student research group on language uses in the family. During the first reflection task, the group defined themselves as plurilingual and stressed that, in their own diverse language uses, language choice and alternations happened for different reasons: for example, they adopted Moroccan Arabic in the group when they did not want the girl from Pakistan to understand something. The group therefore decided to record conversations between three sisters of Moroccan origin at home and between two Pakistani friends. Once the data was collected, they listened to the recordings and reflected on it, in order to systematise what was relevant to transfer onto their digital poster.

Conceptualising communicative resources and orientations (or not) towards the school norm

At the beginning of the project, in order to launch the guiding questions, the teacher evoked the myth of the Tower of Babel, according to which the presence of diverse languages in the world is a punishment for humanity, since it can hinder communication between people. This idea is, in fact, coherent both with the organisation of language education in the high school – different languages are taught separately – and with teaching practices that proscribe more than one language at a time, as we have mentioned before.

This compartmentalised language teaching was evoked by students in tasks previous to the project, as well as in their first reflections on their communicative practices. On the one hand, they commented that languages are used in a given place or with a given person, and on the other hand they did not contemplate their different regional language varieties, or the ones they used when interacting through new technologies. For example, students remarked that they speak Arabic, but did not distinguish between Darija and Standard Arabic, or they mentioned only the languages they consider official and dismissed hybrid varieties, such as the mixed Moroccan Arabic and Spanish resources they often use.

During the development of the project, it was observed how the students oriented towards the school norm according to classroom regimes. However, despite that, sometimes the data students collected showed that the use of various linguistic resources, instead of being compartmentalised, constitutes a continuum of the resources at hand, without clear borders. This is observed in Excerpt 11.2, collected in a Catalan class by the group working on language brokering. Kira starts by introducing the activity.

Excerpt 11.2 Participants: Kira (KIR, Pakistan), Rania (RAN, Bangladesh), Pep (PEP, teacher)

01 KIR	ara us explicarem una miqueta com funcionen les le-
	now we will explain to you how the (translation) works
02	(0.2)
03 KIR	ai que he empezado en catalán
	oh i have started in catalan
04 PEP	no [tu pa- tu pots fer-ho en castellà&
	no [you spe- you can do it in spanish&
05 RAN	[jo:pe:
	[sh:it:
06 PEP	&si és classe de castellà [després jo parlaré en català
	&if it is the spanish class [after i will speak in catalan

Kira begins her presentation in Catalan, orienting herself using the medium of this class. In line 3, she starts a repair of her language choice when she realises that she is collecting data for Spanish class. This is followed by the teacher's alignment to her repair, saying Kira can speak in Spanish. This is not an isolated case. According to our data, Spanish and Catalan form a continuum; resources belonging to different systems may go unnoticed.

After the reflection tasks proposed in the project, the students reached one of their conclusions: that they mix languages or use mixed varieties. This conclusion broke with their above-mentioned initial conceptualisations of their language uses.

To provide evidence about their real language uses, Ousira, a student of Moroccan origin from the group who studied the language uses in the family, recorded a conversation with her two sisters about exams and school.

Excerpt 11.3 Participants: Oussira (OUS), Fatiha (FAT), Hadija (HAD)

01 HAD	yo no yo no tengo exámenes	
	i don't i don't have exams	
02 OUS	tu como siempre hadija tu porque estás en primaria pero ya verás	
	you as usual hadija you because you are in primary but you will see	
03	cuando vas a ir a la eso (0.9) vas a (0.1) [flipa:r]	
	when you will move on to seconday education (0.9) you will (0.1) [flip out:]	
04 FAT	[3ndik_] (.) 3at flipar	
	[i have_] (.) flip out	
05	(0.8)	
06 OUS	n 3tchof (0.8) ki kolak 3aklak o sea daba makat kitahaja (0.7) pero	
	you will see (0.8) you think that that is now you do nothing (0.7) but	
07	ya verás (3.4) o sea ya te puedes ir es[forzando má::s	
	you will see (3.4) that is you should start putting more e [ffort into it	
08 HAD	[que guapa que es esa también	
	[this one is also so pretty	
	((refering to a girl on the tv they are watching))	
09	(0.2)	
10 OUS	ya te puedes ir esforzando más porque si no (1.1) mizian	
	you should start putting more effort into it otherwise (1.1) you will see	
11	(0.9)	
12 FAT	((laughing)) mizian	
	((laughing)) you will see	
13 OUS	((laughing))	

The student research group investigating family language uses conceptualised them as being represented by a Spanish-Arabic continuum. The same applied in the case of Urdu and Spanish. In this way, students declared that, unlike what they indicated in their initial reflections, a place or a person might not imply the use of a single language.

Discursive Outcomes

The digital poster

The first of the discursive products resulting from the project was the digital poster, which turned out to be truly productive since its multimodal affordances allowed the inclusion of different text genres, images, audio and video recordings, integrated coherently in the document. For this reason, students were guided to make their poster in different stages and were helped through specific activities and different mediating artefacts (e.g. observe the characteristics of similar posters; plan the organisation of their poster; write the content of each part; review their work). To construct the digital poster, the students had to organize all the collected data, discuss it and write their analysis to transmit the results. Therefore, the resulting posters were plurilingual, because of the data presented, and multi-genre since, in order to present their research to a public, students needed descriptive texts – to explain, for example, the process of data collection – argumentative texts – to discuss possible reasons for their code-switching – dialogic texts – to portray their real practices – and expositive texts – in which to present facts about their linguistic practices.

Moreover, during the process of doing the project, the students used their plurilingual repertoires, which, we argue, should not be an obstacle to present a final multimodal product in a unilingual mode, following the demands of the school curriculum. Thus, students' plurilingual resources were the medium for working as a group and, in turn, the medium for developing new competences, including those that require the use of a single language for formal communication.

All in all, the digital poster's structure helped the students to form a plurilingual multi-genre discursive outcome useful for sociolinguistic research dissemination. Image 11.1 is an example of this final product.

Presenting the results

One of the characteristics of project-work, and what makes it meaningful, is that the final product is real and transferable to third parties. As we have said, the poster had to be presented to a class of trainee teachers at the university. Thus, the oral presentation in front

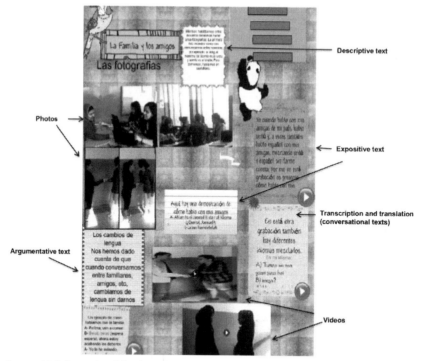

Image 11.1 Example of final product

of an informed audience was the second product resulting from the project, and it involved two relevant aspects: (i) the construction of oral discourse organised following certain guidelines; and (ii) the categorisation of the high school student-researchers as experts in language use.

The student-researchers planned the oral presentation following, on the one hand, the instructions set by the teacher and the researcher: it had to contain a presentation of the group members, an introduction to the topic, a data presentation and analysis, and the most relevant conclusions. Furthermore, the evaluation criteria took into account the timing of the presentation, organisation (content and sequencing), and the overall quality of the students' performance. Video recordings of the presentations show that students achieved the proposed objective of carrying out a quality oral presentation in a unilingual mode, thus meeting the requirements of a formal social practice such as this one.

The groups of student-researchers were able to answer the questions posed by the university students, as we can observe in the next excerpt when the group that worked on family practices answered a question about the differences between standard Arabic and Darija.

Excerpt 11.4 Participants: Oussira (OUS), Jafar (JAF), Naila (NAI)

01 OUS también por ejemplo como yo soy de tanger y él es de marraqueix hay
also for example since i am from tanger and he is from marrakesh there are

02 palabras (0.2) poquitas pero: diferentes sabes/
words (0.2) not a lot of them but: diferent you know/

03 JAF es [como en castellano
it is [the same in spanish

04 NAI [son dialectos
[they are dialects

05 OUS pero yo puedo entender pero no hay veces que no sabes/ hay veces que
but i can understand but no sometimes i can't you know/ sometimes

06 no se puede entender muy bien
it is not possible to understand

07 JAF es como el catalán de barcelona con el de valencia
it is like the catalan from barcelona and from valencia

Ousira informs the university students that there are differences in Darija according to the area. Then Jafar offers a comparison with variation in Spanish. In turn, Naila specifies the term 'dialects' to make reference to the sociolinguistic situation in Morocco. In lines 5–6, Ousira focuses on the difficulties of understanding faced by people from different regions. Finally, Jafar (line 7) establishes a parallelism between what happens in Morocco and in Catalan-speaking regions. The students are thus able to reflect on specific sociolinguistic situations and make comparisons.

In the same way, the students-researchers offered a metalinguistic explanation about 'Arabic chat language' used with mobile phones or Facebook: they discussed its form, but also how they learned it and, therefore, its mode of transmission. These rich reflections were possible thanks to a previous task of producing transcripts of audio recordings in Arabic for which they used this 'chat language'.

Teacher and student evaluations of the project

During the project development, the transformation of the teacher's vision was clearly observed in relation to her conception of plurilingual practices and how to teach formal skills to students with such varied competences. The digital poster constituted a boundary object that was the focus of reflection for improving future projects.

This positive perception of the project outcomes was also reflected in the feedback collected from the students, since their plurilingualism was valued as an academic topic and they considered themselves bearers of knowledge to be shared.

Final Remarks

It is no secret that a monolingual vision of teaching languages prevails. In fact, for decades, plurilingual practices have been understood as evidence of a lack of competence. However, this ideology is beginning to change, both theoretically and in teaching practices, especially thanks to discussions around concepts such as translanguaging and plurilingual practices. While the first emphasises transgression in the face of monolingual ideologies and the empowerment of students by giving value to their repertoire, the second highlights the unavoidable value of whole repertoires for developing new communicative resources through talk-in-interaction.

The fieldwork carried out at the high school in Barcelona reported on in this chapter showed that certain teachers were more accepting of plurilingual practices in the classroom and promoted them to encourage students' participation, taking students' previous competences as the basis for their learning and giving value to them. This previous finding inspired us to systematically develop aspects of the curriculum, such as those relating to linguistic diversity and particular discursive skills.

This was possible by adopting an approach placing students as agents of their learning, proposing a meaningful didactic project, taking into account students' sociolinguistic backgrounds, and empowering them by converting their research into academic knowledge that could be taught to other people. Along the way, and by making use of their whole linguistic repertoires, students engaged with the formal communication skills expected at their level of schooling. According to the results, this didactic proposal afforded transformative potential both for the students – who were able to participate using their whole repertoires and build knowledge and outcomes that met curricular demands – and for the teacher – who changed her monolingual vision about language learning.

Notes

1. Transcription conventions:

Pseudonym of participant	ABC
Overlapping	[
Length of the pause	()
Comments	((laughing))
Rising intonation	/
Falling intonation	\
Lengthening of sound	: ::
Abrupt cut off	-
Turn continues	&
Translation	**translation**

References

Burns, A. (1999) *Collaborative Action Research for English Language Teachers*. Cambridge: Cambridge University Press.

Candelier, M. (ed.) (2007) *Cadre de Référence pour les Approches Plurielles des Langues et Des Cultures* (CARAP). ECML, Conseil de l'Europe. https://carap.ecml.at/Portals/11/documents/C4pub2007F_20080228_FINAL.pdf (accessed January 2020).

Corona, V., Nussbaum, L. and Unamuno, V. (2013) The emergence of new linguistic repertoires among Barcelona's youth of Latin American origin. *International Journal of Bilingual Education and Bilingualism* 16 (2), 182–194

Duff, P. and Kobayashi, M. (2010) The intersection of social, cognitive, and cultural processes in language learning: A second language socialization approach. In R. Batstone (ed.) *Sociocognitive Perspectives on Language Use and Language Learning* (pp. 75–93). Oxford: Oxford University Press.

Duverger, J. (2007) *Didactiser l'alternance des langues en cours de DNL*. *Tréma* 28. http://trema.revues.org/302 (accessed January 2020).

Egan-Robertson, A. and Bloome, D. (eds) (1998) *Students as Researchers of Culture and Language in their own Communities*. Cresskill, NJ: Hampton Press, Inc.

European Council (2001) *Common European Framework of Reference for Languages: Learning, Teaching, Assessment*. Strasbourg: Council of Europe.

European Council (2018) *Common European Framework of Reference for Languages: Learning, Teaching, Assessment. Companion Volume with New Descriptors*. Strasbourg: Council of Europe.

Gafaranga, J. and Torras i Calvo, M.C. (2001) Language versus medium in the study of bilingual conversation. *International Journal of Bilingualism* 5 (2), 195–219.

García, O. (2009) *Bilingual Education in the 21st Century: A Global Perspective*. Malden, MA and Oxford: Basil/Blackwell.

García. O. and Sylvan, E. (2011) Pedagogies and practices in multilingual classrooms: Singularities in pluralities. *The Modern Language Journal* 95 (3), 385–400.

Hall, J.K. (1993) The role of oral practices in the accomplishement of our everyday lives: The sociocultural dimensions of interaction with implications for the learning of another language. *Applied Linguistics* 14 (2), 145–166.

Hall, J.K., Hellermann, J. and Pekarek-Doehler, S. (eds) (2011) *L2 Interactional Competence and Development*. Bristol: Multilingual Matters.

Hellermann, J. (2007) The development of practices for action in classroom dyadic interaction: Focus on task openings. *Modern Language Journal* 91 (1), 83–96.

Idescat (2014) *Població Estrangera. Districtes. 2014* www.idescat.cat/poblacioestrangera/?res=e19&nac=a&b=10 (accessed January 2020),

Lambert, P. (2012) Identifier la pluralité des ressources des élèves en contexte monolingue et normatif. Une enquête ethnographique auprès de lycéennes. In M. Dreyfus and J. Prieurs (eds) *Hétérogénéité et Variation. Perspectives Sociolinguistiques, Didactiques et Anthropologiques* (pp. 182–195). Paris: Michel Houdiard éditeur.

Lave, J. and Wegner, E. (1991) *Situated Learning: Legitimate Peripherical Participation*. New York: Cambridge University Press.

Li, W. (2011) Moment analysis and translanguaging space: Discursive construction of identities by multilingual Chinese youth in Britain. *Journal of Pragmatics* 43, 1222–1235.

Llompart-Esbert, J. (2016) Pràctiques Plurilingües d'Escolars d'un Institut Superdivers: De la Recerca a l'Acció Educativa. Doctoral Thesis. Bellaterra: Universitat Autònoma de Barcelona.

Llompart-Esbert, J. (2017) *La transmissió lingüística intergeneracional inversa: quan fills i filles ensenyen llengua als progenitors*. Treballs de Sociolingüística Catalana 27. www.raco.cat/index.php/TSC/article/view/330444 (accessed January 2020).

Llompart-Esbert, J. and Nussbaum, L. (2018) Doing plurilingualism at school: Key concepts and perspectives. In S. Melo Pfeifer and C. Helmchen (eds) *Plurilingual Literacy Practices* (pp. 19–39). Bern: Peter Lang.

Lüdi, G. and Py, B. (1986/2003) *Être Bilingue*. Bern: Peter Lang.

Lüdi, G. and Py, B. (2009) To be or not to be... a plurilingual speaker. *International Journal of Multilingualism* 6 (2), 154–167

Masats, D., Nussbaum, L. and Unamuno, V. (2007) When activity shapes the repertoire of second language learners. *EUROSLA Yearbook* 7, 121–247.

Masats, D. and Noguerol, A. (2016) Proyectos lingüísticos de centro y currículo. In D. Masats and L. Nussbaum (eds) *Enseñanza y Aprendizaje de las Lenguas Extranjeras en Educación Secundaria Obligatoria* (pp. 59–84). Madrid: Síntesis.

Mondada, L. and Pekarek-Doehler, S. (2004) Second language acquisition as situated practice: Task accomplishment in the French second language classroom. *The Modern Language Journal* 88 (4), 501–518.

Moore, E. and Nussbaum, L. (2016) Plurilingüismo en la formación del alumnado de la ESO. In D. Masats and L. Nussbaum (ed) *Enseñanza y Aprendizaje de las Lenguas Extranjeras en Educación Secundaria Obligatoria* (pp. 15–33). Madrid: Síntesis.

Moore, E., Ploettner, J. and Deal, M. (2015) Professional collaboration at the boundaries between content and language teaching. *Ibérica* 30, 85–104.

Nussbaum, L. (2013) Interrogations didactiques sur l'éducation plurilingue. In V. Bigot, A. Bretegnier and M. Vasseur (eds) *Vers Le Plurilinguisme? 20 Ans Après* (pp. 85–93). Paris: Albin Michel.

Nussbaum, L. (2017) Investigar con docentes. In E. Moore and M. Dooly (eds) *Qualitative Approaches to Research on Plurilingual Education / Enfocaments Qualitatius per a la Recerca en Educació Plurilingüe / Enfoques Cualitativos para la Investigación en Educación Plurilingüe* (pp. 23–45). Dublin, Ireland/Voillans, France: Research-publishing.net.

Nussbaum, L. and Masats, D. (2012) Socialisation langagière en Catalogne: le mutilinguisme comme étayage de pratiques monolingües. In M. Dreyfus and J.M. Prieurs (eds) *Hétérogénéite et Variation. Perspectives Socolinguistiques, Didactiques et Anthropologiques* (pp. 155–167). Paris: Michel Houdiard.

Nussbaum, L. and Rocha P. (2008) L'organisation sociale de l'apprentissage dans une approche par projet. *Babylonia* 3, 52–55.

Nussbaum, L. and Unamuno, V. (2006) *Usos i Competències Multilingües entre Escolars d'Origen Immigrant*. Bellaterra: Servei de Publicacions de la UAB.

Patiño, A. and Unamuno, V. (2017) Investigar prácticas de socialización lingüística de alumnado de origen inmigrante a través de su participación como investigadores. In E. Moore and M. Dooly (eds) *Qualitative Approaches to Research on Plurilingual Education / Enfocaments Qualitatius per a la Recerca en Educació Plurilingüe / Enfoques Cualitativos para la Investigación en Educación Plurilingüe* (pp. 107–128). Dublin, Ireland/Voillans, France: Research-publishing.net.

Pekarek-Doehler, S. (2011) Desmitificar las competencias: hacia una práctica ecológica de la evaluación. In C. Escobar and L. Nussbaum (eds) *Aprendre en una Altra Llengua* (pp. 35–51). Bellaterra: Servei de publicacions de la Universitat Autònoma de Barcelona.

Perrenoud, P. (1999) *Apprendre à l'École à Travers des Projets: Pourquoi? Comment?* Faculté de Psychologie et des Sciences de l'Éducation, Université de Genève. www.unige.ch/fapse/SSE/teachers/perrenoud/php_main/php_1999/1999_17.html (accessed January 2020).

Rodriguez, L. and Brown, T. (2009) From voice to agency: Guiding principles for participatory action research with youth. *New Directions for Youth Development* 123. Wiley Periodicals, INC.

Star, S.L. and Griesemer, J.R. (1989) Institutional ecology, translations, and boundary objects: Amateurs and professionals in Berkeley's museum of vertebrate zoology. *Social Studies of Science* 19 (3), 387–420.

Stetsenko, A. (2014) Transformative activist stance for education: The challenge of inventing the future in moving beyond the status quo. In T. Corcoran (ed.) *Psychology in Education* (pp. 181–198). Rotterdam/Boston/Taipei: Sense Publishers.

Swain, M. and Lapkin, S. (2000) Task-based second language learning: The uses of the first language. *Language Teaching Research* 4 (3), 251–274.

Tse, L. (1996) Language brokering in linguistic minority communities: The case of Chinese- and Vietnamese-American students. *Bilingual Research Journal* 20 (3–4), 485–498.

Wenger, E. (1998) Communities of practice: Learning, meaning and identity. *Journal of Mathematics Teacher Education* 6 (2), 185–194.

Young, R.F. (2007) Language learning and teaching as discursive practice. In Zhu H., P. Seedhouse, W. Li and V. Cook (eds) *Language Learning and Teaching as Social Inter-Action* (pp. 251–271). Basingstoke: Palgrave Macmillan.

12 Translanguaging as Practice and as Outcome: Bridging across Educational Milieus through a Collaborative Service-Learning Project

Claudia Vallejo Rubinstein

Introduction

One of the biggest challenges of translanguaging as a transformative pedagogy continues to be its acceptance by educators as being a valuable and legitimate resource *per se*, beyond the use of translanguaging as scaffolding to achieve learning outputs in a standard academic target language (García & Li, 2014: 125). That is, the challenge is to acknowledge and promote translanguaging as an asset that all children, teachers and trainee teachers should master. Understanding translanguaging as 'a form of communication that relies on a willingness to engage in communicative practice which blurs or breaks through apparent boundaries between languages, signs, codes, and cultures' (Blackledge & Creese, 2016), a translanguaging pedagogy could help to develop pupils' creativity, their plurilingual and pluriliteracy competences (García *et al.*, 2007), and to promote learning spaces that are more inclusive of pupils' everyday communicative practices.

In this chapter I will describe a project that builds on translanguaging as practice and outcome while bridging across two educational milieus: an after-school literacy program in a multicultural, multilingual primary school in Barcelona for children classified as being 'at-risk' of not meeting established curricular objectives, and a service-learning project where trainee teachers at a Catalan university developed educational resources incorporating a translanguaging approach for – and with – the after-school program participants (see Moore & Vallejo, 2018, for a discussion on the notion of 'at risk' students as it is used in this research context). The project aimed at incorporating the everyday languaging practices of plurilingual students, and their voices as creative agents, into the legitimate practices of the after-school program, and into the development and everyday praxis of future classroom teachers, therefore linking informal and mainstream educational contexts. The aim was to bridge between translanguaging in social life, and translanguaging as pedagogical theory and practice. In this sense, this project aligns with others described in this volume – such as Bradley and Atkinson's, King and Bigelow's, and Llompart-Esbert and Nussbaum's contributions – which explore how collaborative work might open new possibilities for translanguaging, including how it might be brought into the classroom.

In the following sections I will present the context and some examples of the ethnographic work carried out at the after-school program, before describing the collaborative service-learning project and its outcomes. Prior to this, though, I will define the main theoretical principles that inform this research and their relation to translanguaging.

On Plurilingualism, Pluriliteracies and Translanguaging

Plurilingualism 'presupposes the existence of a free and active subject who has amassed a repertoire of resources and who activates this repertoire according to his/her need, knowledge or whims, modifying or combining them where necessary' (Lüdi & Py, 2009: 157). Studies on plurilingualism draw on a long European (especially Francophone) research tradition, including the work of interactional sociolinguists who build on Gumperz' (1964) conceptualisation of a speaker's repertoire as a unique and integrated set of semiotic resources for communication, and focus on speakers' languaging practices in interaction, rather than on languages as discrete systems. Such key principles are also at the basis of translanguaging (García, 2009; see also García & Otheguy, 2019).

The term pluriliteracies, as proposed by García et al. (2007) builds on previous literature on bi- and multiliteracies, adding the principles of plurilingualism to develop a perspective that accounts for flexible, fluid literacy practices around plurilingual, intercultural and multimodal texts where the individual's full communicative repertoire is displayed. In this sense, pluriliteracies refers to a holistic approach to the literacy practices of plurilingual people and conceptualises reading, writing and other activities engaged around text as involving much more than mastering linguistic coding and decoding skills (Moore & Palou, 2018: 79).

In a more recent theoretical development, Hawkins (2018: 55) describes the 'trans-' turn in language and communication studies – led in part by research on translanguaging – to refer to 'the current era of globalization in which communication occurs with ever-increasing rapidity among ever-expanding audiences, through rapidly changing semiotic means and modes'. According to Hawkins (2018), this 'trans-' turn also highlights 'the significant increase of attention to the ways in which language is enmeshed with other semiotic resources in constructing meanings in communication' in fluid and unpredictable ways. The 'trans' prefix also accounts for a deep commitment towards social and educational transformation, understood here as destabilising and expanding the institutional norms and the practices typically valued in academic contexts, towards educational practices that are more inclusive of the uses and competences of plurilingual language users in the 21st century.

Despite their nuances, in this text, plurilingualism, pluriliteracies and translanguaging are conceived as complementary. As Moore and Vallejo state, in discussing the same research context as the one presented in this chapter:

> While the notions of plurilingualism and pluriliteracies are useful for describing the repertoires of the children with whom we work, the notion of translanguaging is useful to conceptualise how these repertoires might be mobilised to transgress and eventually transform practices and structures that are dominant to educational systems, such as those favouring mono-competences. (2018: 27)

Context

The collaborative study described herein is part of a larger, three year (2014–2017) linguistic ethnography in an after-school reading program. The initiative is supported by volunteer mentors and is aimed at 4th and 5th grade primary school students (9–10 years old) considered 'at-risk' of not meeting established curricular objectives. The program is run by a non-profit organisation in several schools and libraries in Catalonia. The specific site of this ethnographic work was a public primary school in a socioeconomically disadvantaged neighbourhood in Barcelona. According to the school's head teacher, 90% of the school population had immigrant backgrounds. Half of these children were born in Catalonia and half abroad, in some cases arriving at the school with previous formal education in their countries of origin. The composition of the after-school program during the ethnographic work reflected that of the school: 22 of the 24 participating children had immigrant family backgrounds, as well as highly plurilingual repertoires including home languages different from those of the school curriculum (which includes Catalan as the vehicular language, Spanish as a second language, and English as a foreign language). Despite this diversity, neither the coordinators nor the volunteers had much insight into the children's linguistic repertoires and everyday language practices.

The main objectives of the after-school program are to promote the children's love for reading and develop their reading habits, in order to improve their educational opportunities. The main dynamics are reading one-on-one with an adult volunteer and playing literacy-related games, such as hangman and riddles. Although there are no explicit norms regarding language use, all the reading materials and related activities provided by the program are in Catalan, and volunteers encourage children to communicate and read in that language.

Despite this foregrounding of the primary language of the children's schooling, the ethnographic research reveals that the program is also a space where participants can engage in fluid practices and display competences that might otherwise remain invisible in formal schooling. Although these do not seem to be deliberate aims of the program, plurilingual children blur socially constructed boundaries between named languages and between academic and everyday language practices, thus opening up non-official translanguaging spaces (Li, 2011, 2017) that encompass their whole communicative and cultural repertoires in creative and transformative ways (for more details on the ecology of the after-school program, see Moore & Vallejo, 2018; Vallejo & Moore, 2016). These observations align with previous research in non-formal educational contexts that highlights the value of non-curricular spaces for being more inclusive and legitimising of children's whole language practices. From a translanguaging perspective, García and Li (2014) describe several cases that 'give evidence of how it is especially in alternative educational spaces that translanguaging as pedagogy often fulfils its promise to truly open up spaces for meaning-making and social justice' (2014: 117), describing these settings as 'more amenable' to the development of translanguaging as critical pedagogy (2014: 90). Other authors (Blackledge & Creese, 2010; Helm & Dabre, 2018; Li, 2011, 2014; among others) highlight the potential of non-formal educational settings to generate collaborative spaces where multimodal practices and what Creese and Blackledge (2010) call 'flexible bilingualism' are praised as beneficial for learning, challenging monolingual ideologies that tend to reify traditional language expert-learner roles and interactions.

Snapshots of Emergent Practices within the After-School Program

Many of the interactions documented during the ethnographic work at the after-school program show the 'multiple discursive practices in which bilinguals engage in order to make sense of their bilingual worlds' (García, 2009: 45). I will briefly describe an example: Hasu is a 10 year old girl born in Indian Punjab who had moved to Barcelona with her family 2.5 years before I met her. In India, she had been schooled in an English-medium school, and in Barcelona, she had recently taken reading and writing lessons in Punjabi, her home language, at the Sikh temple. Hasu and her mentor in the after-school program agreed to borrow a book in Punjabi from the public library and to read one or two pages – which included text in Punjabi and illustrations – at the beginning of each session. In this fragment, Hasu is explaining the content of a page she just read, which includes text in Punjabi and an illustration of a woman cooking and a bird picking some kind of dough from a plate.

Fragment 12.1

(1) [noise from other people in the room]
(2) H: i diu que quan ella està menjant... a la... la casa
(3) H: a... ve sempre ve un això [points at drawing in the book] un... un *sparrow* volant per *la cocina* i ve a seu aquí [points at drawing] i es menja
(4) H: menja la... això [points at another part of drawing]
(5) H: i la seva mare està... quan cuinem en això [points at another part of drawing] el *chapati se infla* [makes a gesture like a growing balloon with both hands]
(6) H: la mare a... està cuinant la... el *chapati* inflat.
(7) M: aha.

Translation

(1) [noise from other people in the room]
(2) H: and says that when she is eating... at... at home
(3) H: a... comes always comes this [points at drawing in the book] a... a *sparrow* [English in the original] flying by *the kitchen* [Spanish in original] and comes seats here [points at drawing] and eats
(4) H: eats the... this [points at another part of drawing]
(5) H: and her mother is... when we cook in this [points at another part of drawing] the *chapati* [indian bread] *expands* [Spanish in

original] [makes a gesture like a growing balloon with both hands]
(6) H: the mother a... is cooking the... the expanded *chapati*.
(7) M: aha.

This fragment shows how a range of complex practices and different modes intertwine to create and communicate meaning during this activity. One of the most salient, which is actually in the origin of the term translanguaging (as coined by Williams, 1996, translated in Baker, 2001) consists of reading in one language and producing an explanation in another, motivated in this case by Hasu's (H) objective of sharing her reading with her mentor (M). But to do so, and because she is an emergent plurilingual whose complex repertoire is under construction and whose mastery of the different features that make up her plurilingual repertoire is not balanced, Hasu dynamically displays resources and uses diverse strategies to navigate across obstacles and overcome what García (2009: 53) describes as the 'ridges and craters' of communication.

Thus, in her explanation, Hasu calls upon her knowledge of a number of languages (Catalan, Spanish, English and Punjabi), while also incorporating other resources such as indexing the images on the page or gesturing, thus engaging in what Baynham *et al.* (2015) describe as interlingual and intersemiotic translanguaging. At some points, she interrupts her explanation to introduce specific cultural information (preparation of the chapati) that her mentor might not know and needs to understand the action being described, thus also culturally and contextually orienting her production to her interlocutor and bridging between languages and cultures.

Hasu is not shy about using her entire repertoire and all her available resources to make and communicate meaning. Although the majority of her production is in Catalan, it is not clear that she is actually orienting to the production of a unilingual (or 'one language at a time', see Llompart-Esbert & Nussbaum in this volume) explanation in the school's vehicular language, as she displays resources from other languages and modes without requesting translations from her mentor. In fact, her translanguaging is not negatively sanctioned or repaired by herself or others in the interaction, but rather validated by her and her interlocutor's actions. In this sense, translanguaging appears not just as a scaffolding strategy, but as a legitimate and significant resource which allows the task to move on.

From a normative and monolingual understanding of language and literacy, what Hasu does could be seen as signalling 'deficit' as she fails to produce unilingual speech in the vehicular language of the school and of the after-school program. However, from a translanguaging (and plurilingualism or pluriliteracies) perspective that puts fluidity and social action at the forefront, we might focus on her display of competence in engaging in complex plurilingual and multimodal interactional activities around reading to achieve effective communication. Her competence is displayed by drawing on many different modalities and resources that complement, compensate and reinforce each other, while orienting her production to her interlocutor. Hasu builds meaning within, between and beyond named languages, bridging, expanding and transgressing borders between languages and other semiotic systems, and between signs, codes and cultures (Blackledge & Creese, 2016). By doing so, she positions herself and is positioned as a competent plurilingual reader and communicator, a role afforded by the flexible distribution of agency, expertise and competence between the child and the mentor.

Many similar snapshots of interactions within the program illustrate the practices of translanguaging or of pluriliteracies that emerge when children display agency in selecting materials and activities, and draw on their full communicative repertoires. Still, the systematic observation of the program sessions has allowed a certain distance and oftentimes tension to be identified between the hybrid and fluid languaging that children display and the program's and mentors' more normative understanding of language and literacy based on monolingual ideologies. These tensions might reflect the coexistence of approaches to language use and language learning based on a strict separation of language systems, and more fluid conceptualisations of languaging as practical social action.

Intertwined with this, tensions also emerge around the formats, genres and activities that volunteers consider legitimate for literacy support, which are usually monomodal and monolingual, and the children's preferences, which usually include multimodal and plurilingual activities around comics, Japanese manga, computer games and musical lyrics and videos. In most of the interactions documented and analysed these tensions were solved by children and volunteers negotiating activities and displaying flexible and interchangeable roles.

By engaging in such practices, Hasu and other children from the program, as well as their mentors transcend traditional school literary activities towards pluriliteracies, displaying plurilingual and multimodal resources to communicate, to learn (including the learning of Catalan) and to collaboratively make meaning.

Hasu and her peers were placed in the program due to being categorised by school agents as deficient in terms of literacy, although their interactional practices bring this categorisation into question. Hasu's ability to lead the reading activity in a highly competent way and be understood by her mentor was afforded by her freedom to choose and to structure the reading practice, drawing on diverse linguistic and semiotic resources according to her interests, experiences and competences (see Moore & Vallejo, 2018, for a similar argument).

Many similar cases were documented during the ethnographic work, where children actively sought and negotiated conditions to display competences not usually validated in formal educational environments, practices that allowed them to position themselves and be positioned as active agents and competent readers and language users (Moore & Vallejo, 2018; Vallejo & Moore, 2016). These cases correspond to what has been called natural translanguaging (Williams, 2012) or pupil-directed translanguaging (García & Li, 2014; Lewis *et al.*, 2012), since they emerged in the participants' emergent practices but were not part of a deliberate pedagogical approach for developing literacy.

By documenting pupil's natural plurilingual and pluriliteracy practices through the ethnographic research, interest emerged in collaboratively producing pedagogical resources and structured activities that, drawing on these practices, enhanced, validated and opened spaces for translanguaging. A collaborative project including university students in a service-learning endeavour was thus implemented, and will be described in the following section.

Translanguaging as Pedagogy for Pluriliteracies

The collaborative service-learning project

One of the main aims of the collaborative project described herein was to contribute to the after-school program's configuration as an officially available, creative, 'in-between' space where children and volunteers would collaboratively articulate meaningful and flexible practices for the learning and enjoyment of literacy, rather than perpetuating common dichotomies of what counts or not as valid, or what can and cannot be done in and out of school.

This rationale links to Gumperz and Cook-Gumperz's (2005) research and advocacy on opening educational spaces to students' fluent language practices, thus creating a safe 'interactional space in which students are free to use all their bilingual resources' (2005: 1) and collaboratively develop their plurilingual competence. It also relates to Li's (2011) definition of a translanguaging space (2011: 1223), a space created 'by and for translanguaging practices', where language users break social and ideological boundaries between named languages, varieties and other semiotic resources, and which allows them to 'integrate linguistic codes that have been formerly separated through different practices and in different places'. From the perspective of literacy, it also links to Gutierrez's (2008) definition of the development of critical literacy (2008: 149), described as 'contesting traditional conceptions of academic literacy and replacing / introducing forms of literacy that privilege and are contingent upon students' sociohistorical lives, including hybrid language practices, play and imagination'.

Many of these features were already present in the children's practices and the child-volunteer interactions, but did not find resonance in the program's structure, offer of materials and official dynamics. The aim was to build on pupil-directed plurilingual practices to promote an institutional translanguaging space where all agents could engage in critical pluriliteracies through explicit translanguaging pedagogies. For articulating such a project, future Primary Education teachers from a subject called School Language Project and Plurilingualism at the Catalan university where the author lectures were invited to develop learning resources, while all the program's agents agreed to actively contribute to their production and testing. These resources, conceptualised in the second year of the research as 'plurilingual challenges' (for a discussion of the service-learning project in the first year see Moore & Vallejo, 2018), encouraged children and volunteers to draw on their entire repertoires for making meaning and for learning, including the learning of the school's vehicular language. They also expanded the boundaries of the program by incorporating children's families and home language practices, bringing in new actors and contexts.

The introduction of trainee teachers into the equation allowed the project to expand towards new objectives: preparing future teachers for our cosmopolitan societies by incorporating a translanguaging approach in their training and praxis, and engaging in teaching and learning experiences across diverse educational milieus, overcoming the current situation where 'learning activities that take place in and out of school are often not mutually recognized' (Subero et al., 2017: 247). In this sense, this project aligns with others that explore the role of after-school programs in bridging across different social and cultural milieus, including schools, families and minority communities (Crespo et al., 1999, 2005; Hawkins, this volume; Lee & Hawkins, 2008; Subero et al., 2017). By bridging between formal, non-formal and family settings, the project also aimed at promoting what García et al. (2013: 814) describe as transcollaboration: the creation of 'extended learning support communities beyond the traditional school structure'.

This interest in building extended support communities and working collaboratively imprinted the whole process: volunteers and children from the after-school program acted both as experts/advisors and evaluators of the trainee teachers' proposals, in an interactive process where all participants contributed to the outcomes while displaying fluid roles. This transcollaborative network was designed under the premises of service-learning, a proposal that aims to create cooperative social networks by blending pedagogical and community service objectives. Service-learning builds on a Vygotskian sociocultural approach to learning as socially situated and collaboratively produced, on Freirean critical pedagogy and on Deweyan democratic, participatory educational approaches. Its principles include a critical transformative approach to education and the placing of social justice and social inequalities at the forefront of educational agendas. Documented experiences of service-learning highlight that it enhances classroom learning through practical involvement in issues discussed in class, brings new elements to the curriculum, links learning to real life, and empowers students as active agents of their learning process and of social change (Billig, 2000; Fundació Jaume Bofill, 2015; Kinloch & Smagorinsky, 2014).

Involving trainee teachers in a collaborative service-learning project was an opportunity for them to learn about this approach, as well as to be introduced to ways of connecting non-formal educational settings, which are often absent from their academic development, and mainstream education. More specific objectives included engaging with a real after-school literacy support program in a disadvantaged community and working collaboratively with other educational agents; documenting and valuing the plurilingual resources and practices brought by the children and their families; and applying theoretical concepts worked on in class to the collaborative development of children's pluriliteracies.

Translanguaging was both an objective and a central feature of the process: while the subject of School Language Project and Plurilingualism was delivered mainly in English, trainee teachers had to create their proposals in English and Catalan, while also incorporating the children's home and other languages, some unknown to them, for which they had to do research and collect information.

Specific circumstances did not allow the university students to physically attend the program's sessions, as they had classes at the faculty while it was running, and also because the presence of over 70 university students would have been disruptive to the program's one-to-one dynamics. To compensate for these constraints, special emphasis was put on providing data and support from the after-school program's agents to allow a deep understanding of the program's dynamics. Consequently, the first step was thoroughly presenting students with the ecology of the program as documented in the ethnographic research (as described in Moore & Vallejo, 2018; Vallejo & Moore, 2016). Multimodal data was presented for them to analyse, including videos and transcriptions of child-volunteer interactions and program materials. Also, based on the researchers' field notes, a selection of case studies in the form of vignettes were produced as teaching materials and shared with the trainee teachers, to help them become familiar with the dynamics of the program. Students were then invited to analyse and reflect before embarking on the production of the 'plurilingual challenges' in small groups, acknowledging the importance of developing translanguaging pedagogical resources 'ground up, from the practices we see multilingual students adopting' (Canagarajah, 2011: 415).

Furthermore, a volunteer from the program (referred to here as Pina, a pseudonym) was invited to the university as an expert source of information. Since the beginning of her participation in the program, Pina had expressed an interest in pursuing ways of integrating children's plurilingual and pluriliteracy practices into the reading sessions, and was especially motivated and creative in preparing meaningful materials and activities that encouraged the child she read with to display her full repertoire. Her contribution to the university students' understanding of the program was especially relevant due to the physical constraints mentioned above. Trainee teachers presented their doubts and initial proposals to Pina, who helped them design materials according to the specific conditions and affordances of the program's sessions. These needed to be engaging, playful and meaningful activities, that allowed participants to draw on their full communicative repertoires and multiple cultural experiences, and which could be developed in one-hour sessions, in child-volunteer pairs or small groups, but also beyond the sessions with their friends and family.

After reviewing their initial proposals with Pina, the trainee teachers incorporated her feedback to produce improved versions of around 20 sets of materials which were then compiled into a single document and presented to the children and volunteers to test and evaluate. They were asked to go through the activities, select the ones they would like to do, and give feedback by filling out an evaluation form which also collected information on children's language knowledge and practices. With these forms in hand, the trainee teachers selected the most positively evaluated activities, incorporated the participants' feedback and reshaped them into multimodal flashcards with clear visual instructions. These were then compiled into a box-set under the title *Juguem amb les llengües!* (Let's play with languages!). These materials were then presented to the program participants and became part of the regular resources available for children to select in every session. All in all, the whole process involved the various agents in several collaborative cycles of production, testing, evaluation, reflection and revision.

Many pedagogical proposals to promote translanguaging and pluriliteracies have been designed within the project, and it would be impossible to include them all here. The following table contains examples of some of the activities proposed:

Table 12.1 Examples of 'plurilingual challenges' proposed by the trainee teachers

Activities
Production and reading of plurilingual and multimodal texts Creating plurilingual comics, Christmas songs, etc.
Plurilingual gymkhana Competing in solving a series of plurilingual and multimodal challenges
Plurilingual crosswords and other traditional literacy games incorporating translanguaging
Playing dominos and memory games in which pairs consist of matching concepts/items displayed in different languages and with other semiotic codes (images, symbols)
Collaborative creation of a plurilingual word map
Creating 'the craziest sentence' Creatively combining words from different languages or creating translanguaging word compounds and concepts
Sharing and translating favourite songs across languages
Doing plurilingual versions of favourite songs and audio or video-recording them
Translating traditional stories and riddles across languages with help from parents
Comparing similar texts from different origins and in different languages E.g. which are the main characters in legends and fairy tales from different origins?
Playing with word cognates across languages
Reflecting on the origin of familiar words Gathering loanwords with help from parents
Documenting school and/or neighbourhood linguistic landscapes
Documenting linguistic diversity on food labels by taking pictures at home
Producing multimodal language autobiographies to document and make visible linguistic repertoires, language uses and communicative skills
Bringing favourite literacy materials from home to present to their mentors
Producing bilingual magnets to help parents do the shopping list and shop for groceries
Recommending favourite readings, movies or computer games to mentors and peers
Visiting the local library and documenting the diversity of resources Which languages could they find? Which genres and formats – comic, manga, magazines, DVD, maps… – were available for them to borrow?
Producing a plurilingual and multimodal school diary for their parents describing daily activities

---

Producing a postcard
Describing an imaginary holiday trip, researching the destination to create a
proper postcard description including visual and textual resources, and sending
the postcard to a real partner requesting an answer

This last activity was chosen by a group of four children and two volunteers who decided to jointly create a postcard for two volunteers who were absent that day and deliver it to them in the next session. The whole activity was pupil-directed. First, they discussed their travel destination, agreeing to include all the countries where they or their families came from, including the volunteers' countries. They used a world map to decide the best travel route. While some drew the postcard's front image, consisting of a suitcase with stickers from all their destinations, others checked the map to provide help on the countries' names and the itinerary order. The writing was also a collaborative process where each child wrote one paragraph, the content of which was collaboratively discussed and agreed upon before and during the writing.

While the children drew on multiple semiotic resources to negotiate and develop the task through translanguaging, the written text was produced in Catalan, although the instructions said nothing about language choice, proving their ability to move from a plurilingual to a unilingual (or 'one language at a time') mode. This orientation might have been driven by the fact that their intended addressees were two volunteers with whom they usually spoke in Catalan, as well as school and program expectations. Similar to the argument put forward in this volume by Llompart and Nussbaum, according to García and Silvan:

> Translanguaging is not only a way to scaffold instruction and to make sense of learning and language; it is part of the discursive regimes that students in the 21st century must perform, part of a broad linguistic repertoire that includes, at times, the ability to function in the standardized academic languages required in schools. It is thus important to view translanguaging as complex discursive practices that enable bilingual students to also develop and enact standard academic ways of languaging. (2011: 389)

While encompassing their in- and out-of-school experiences, these collaborative and translanguaging practices reflected and reconfigured children's' subjectivities. They used their school and home languages, other semiotic resources and socialisation experiences to collaboratively construct a 'journey' that included them all, while also rendering visible their competence in the production of unilingual texts in line with mainstream academic standards. Somehow, the activity facilitated opening an in-between space where school and out-of-school languaging and literacies could meet and expand in more dialogic and less exclusionary ways, to collaboratively engage children and volunteers in meaningful and empowering activities to learn and to be.

Program coordinators have incorporated these proposals into the materials regularly provided for all the sites where the after-school program runs in downtown Barcelona, and have expressed interest in continuing with the collaboration. Other evidences of transformation include the incorporation of bi- and plurilingual books into the program resources, the recording of an institutional video including the children's family languages – thereby valuing the linguistic diversity of the children as an asset of the program – and the acknowledgement, by children and volunteers, of shared language repertoires that had remained invisible, allowing for new commonalities and singularities to emerge.

Hopefully, this collaborative project will contribute to the promotion of translanguaging pedagogies as a regular element of the program's dynamics, and to the creation of long-term working networks for the benefit of all actors engaged.

Some Final Reflections

The prefix 'trans' in translanguaging refers not only to creatively and critically transgressing social boundaries between languages and with other semiotic systems, but also to deliberately advocating for transformation towards more inclusive and socially just educational approaches that promote more and better opportunities for all students. This transformational aspiration has shaped the understanding of the why, how and for whom of this research and the overall conception of the researchers' praxis as inevitably related to collaboration and activism. This prompted the need to go beyond documenting the fluid language and literacy practices of the after-school program towards collaborating with its agents for the program's improvement.

I acknowledge the limitations of a small-scale project to transform highly extended visions and practices around what counts as valid language and literacy education. In this closing section, I would like to reflect on the real scope and implications of 'transformation', as facilitated by a translanguaging approach, within this project. My claims are in every moment bounded within these specific arrangements and constraints.

Doing linguistic ethnography implies approaching an 'object' constituted by dynamic, socially constructed practices in which the researcher also takes part (Heller, 2008), thus overcoming supposed claims of external neutrality. Moreover, doing ethnographic research on and from a translanguaging perspective implies adopting a transformative activist stance (Vianna & Stetsenko, 2014), placing the commitment to document and fight social and educational inequalities at the centre of the research praxis, understood as a collective and collaborative task. A translanguaging approach implies questioning what counts as valid knowledge, how we position ourselves regarding the research site and participants, how we involve others in the research process, and on the meaning of research impact beyond its traditional allocation in academia. Quoting Ballena *et al.* (this volume), it involves problematising research participants' status and the social distribution of research outputs and profits.

By giving voice and agency to all participants, the collaborative service-learning project discussed in this chapter embodies these principles. Potential social gains of this research include learning with and from the communities that were studied; providing the after-school program with resources to advance in their commitment to help disadvantaged children break out of the category of 'at risk' by opening up spaces for translanguaging as pedagogy; bridging between formal and non-formal educational contexts through collaborative networks; preparing future teachers for our cosmopolitan societies by incorporating a translanguaging approach in their training and praxis; and hopefully, contributing to transform the educational experiences and trajectories of their future students in mainstream education.

Acknowledgements

This chapter has been written within the framework of the PhD in Education program at the Universitat Autònoma de Barcelona [aquest treball ha estat realitzat en el marc del programa de Doctorat en Educació de la Universitat Autònoma de Barcelona], and with support from the project Knowledge for Network-based Education, Cognition and Teaching (KONECT, EDU2013-43932-P). The project described in this text could not have been possible without my co-researcher Emilee Moore's contributions, and the unconditional support from the after-school program coordinators, A. Cosials, S. Cortina and J. Ventura; the volunteers, especially P. Socias; and the children. I finally thank my university students for committing to this project and sharing their proposals with the wider community.

References

Baynham, M., Bradley, J., Callaghan, J., Hanusova, J. and Simpson, J. (2015) Translanguaging business: Unpredictability and precarity in superdiverse inner city Leeds. *Working Papers in Translanguaging and Translation* 4. https://tlang754703143.files.wordpress.com/2018/08/translanguaging-business.pdf (accessed January 2020).

Baker, C. (2001) *Foundations of Bilingual Education and Bilingualism* (3rd edn). Clevedon: Multilingual Matters.

Billig, S. (2000) Research on K12 school-based service learning: The evidence builds. *Phi Delta Kappan* 658.

Blackledge, A. and Creese, A. (2010) *Multilingualism: A Critical Perspective*. London: Continuum.

Blackledge, A. and Creese, A. (2016) Translanguaging as cultural and cosmopolitan competence. Paper presented at *IALIC 2016*. Barcelona, 25–27 November 2016.

Canagarajah, S. (2011) Codemeshing in academic writing: Identifying teaching strategies of translanguaging. *The Modern language Journal* 95, 401–417.

Creese, A. and Blackledge, A. (2010) Translanguaging in the bilingual classroom: A pedagogy for learning and teaching? *The Modern Language Journal* 94 (1), 103–115.

Crespo, I., Pallí, C., Lalueza, J.L. and Luque, M.J. (1999) Intervención educativa, comunidad y cultura gitana. Una experiencia con nuevas tecnologías: la casa de Shere Rom. In M.A. Essomba (ed.) *Construir la Escuela Intercultural. Reflexiones y Propuestas para Trabajar la Diversidad Étnica y Cultural* (pp. 185–194). Barcelona: Graó.

Crespo, I., Lalueza, J.L, Portell, M. and Sanchez Busqués, S. (2005) Communities for intercultural education: Interweaving practices. In M. Nilsson and H. Nocin (eds) *Schools of Tomorrow: Teaching and Learning in Local and Global Communities* (pp. 27–62). Bern: Peter Lang.

García,O. (2009) *Bilingual Education in the 21st Century: A Global Perspective*. West Sussex: Wiley-Blackwell.

García, O., Bartlett, L. and Kleifgen, J. (2007) From biliteracies to pluriliteracies. In P. Auer and W. Li (eds) *Handbook of Multilingualism and Multilingual Communication* (vol. 5) (pp. 207–228). The Hague: Mouton de Gruyter.

García, O. and Silvan, C.E. (2011) Pedagogies and practices in multilingual classrooms: Singularities in pluralities. *The Modern Language Journal* 95 (3), 385–400.

García, O., Homonoff Woodley, H., Flores, N. with Chu, H. (2013) Latino emergent bilingual youth in high schools: Transcaring strategies for academic success. *Urban Education* 48 (6), 798–827.

García, O. and Li W. (2014) *Translanguaging: Language, Bilingualism and Education*. New York: Palgrave Macmillan.

García, O. and Otheguy, R. (2020) Plurilingualism and translanguaging: Commonalities and divergences. *Journal of Bilingual Education and Bilingualism* 23 (1), 17–35, (special issue edited by M. Dooly and C. Vallejo on The Evolution of Language Teaching: Towards Plurilingualism and Translanguaging).

Gumperz, J. (1964) Linguistic and social interaction in two communities. *American Anthropologist* 66 (6), 137–153.

Gumperz, J. and Cook-Gumperz, J. (2005) Making space for bilingual communicative practice. *Intercultural Pragmatics* 2 (1), 1–23.

Gutiérrez, K. (2008) Developing a sociocritical literacy in the third space. *Reading Research Quarterly* 43, 148–164.

Hawkins, M. (2018) Transmodalities and transnational encounters: Fostering critical cosmopolitan relations. *Applied Linguistics* 39 (1), 55–77.

Heller, M. (2008) Doing ethnography. In M. Moyer and Li W. (eds) *Blackwell Guide to Research Methods on Bilingualism* (pp. 249–262). Oxford: Blackwell.

Helm, F. and Dabre, T. (2018) Engineering a 'contact zone' through translanguaging. In M. Dooly and C. Vallejo (eds) Bridging across languages and cultures in everyday lives: New roles for changing scenarios. *Language and Intercultural Communication* 18 (1), 144–159.

Kinloch, V. and Smagorinsky, P. (2014) *Service-Learning in Literacy Education*. Charlotte: Information Age Publishing.

Lee, S.J. and Hawkins, M. (2008) "Family is here": Learning in community-based after-school programs. *Theory into Practice* 47 (1), 51–58.

Lewis, G., Jones, B. and Baker, C. (2012) Tranlanguaging: Developing its conceptualisation and contextualisation. *Educational Research and Evaluation* 18 (7), 655–670.

Li, W. (2011) Moment analysis and translanguaging space: Discoursive construction of identities by multilingual Chinese youth in Britain. *Journal of Pragmatics* 43, 1222–1235.

Li, W. (2014) Who's teaching whom? Co-learning in multilingual classrooms. In S. May (ed.) *The Multilingual Turn: Implications for SLA, TESOL and Bilingual Education* (pp. 167–190). NY: Routledge.

Li, W. (2017) Translanguaging as a practical theory of language. *Applied Linguistics* 39 (1), 9–30.

Lüdi, G. and Py, B. (2009) To be or not to be ... a plurilingual speaker. *International Journal of Multilingualism* 6, 154–167.

Moore, E. and Palou, J. (2018) Reading in multilingual environments. In S. Melo-Pfeifer and C. Helmchen (eds) *Plurilingual Literacy Practices* (pp. 79–101). Bern: Peter Lang.

Moore, E. and Vallejo, C. (2018) Practices of conformity and transgression in an out-of-school reading programme for 'at risk' children. *Linguistics and Education* 43, 25–38.

Subero, D., Vujasinovic, E. and Esteban-Guitart, M. (2017) Mobilising funds of identity in and out of school. *Cambridge Journal of Education* 2, 247–263.

Vallejo, C. and Moore, E. (2016) Prácticas plurilingües 'transgresoras' en un programa extraescolar de refuerzo de la lectura. *Signo y Seña* 29, 33–61.

Vianna, E. and Stetsenko, A. (2014) Research with a transformative activist agenda: Creating the future through education for social change. In J. Vadeboncoeur (ed.) *Learning In and Across Contexts: Reimagining Education* (pp. 575–602). New York: Teachers College, Columbia University.

Williams, C. (1996) Secondary education: Teaching in the bilingual situation. In C. Williams, G. Lewis and C. Baker (eds) *The Language Policy: Taking Stock* (pp. 39–78). UK: CAI.

Williams, C. (2012) *The National Immersion Scheme Guidance for Teachers on Subject Language Threshold: Accelerating the Process of Reaching the Threshold*. Bangor: The Welsh Language Board.

Afterword: Starting from the Other End

Angela Creese

In her Foreword to this volume, and in reference to the favourable direction the chapters in this collection take, Ofelia García points out that 'work here starts from the other end'. In this seemingly innocuous phrase, García critiques structuralism and argues for a radical rethink of a linguistics which over-emphasizes codes, categories, and components, at the expense of relationships, people, and communication. Instead the chapters here, she acknowledges, commence with 'the other end', that is, with people forging collaborations to jointly construct 'new linguistic realities', to challenge a notion of multilingualism as peripheral, marginal, and irrelevant, when in fact it is normal, commonplace, and fundamental to life in the 21st century. For many, up-ending linguistics in favour of anthropology is a clarion call for educationalists, researchers and teachers to rethink how current policies mediated by a too narrowly defined linguistics bring untold harm to those who already have most to lose. García describes how linguistic structuralism has produced a persistent colonialism in which whiteness and monolingualism are normative, and serve to subjugate, disengage, and miseducate minoritized people, and minoritized communities. In her Foreword García stays focused on her enduring ambition to transform mainstream schools, and deliver genuine change. If this book raises any caveats for her, they are in relation to the distraction from this goal. She asks if an arts and aesthetics focus can achieve the criticality necessary to disrupt and 'open up cracks' in schools as deeply iniquitous institutions. As she points out, the role of translanguaging is not solely to scaffold, or teach, but it is a journey of transformation to a new politics that embraces the practices of minoritized lives, 'free of judgement from the white monolingual listening subject'. Her words of caution about the aesthetic and 'transcendental' direction is its possible over-emphasis on the emotional at the expense of the criticality required to challenge the hegemonies of the status quo.

For many reading this book there will be satisfaction in finding a space for considering the ways in which applied linguists can engage in different forms of collaboration, including with those in the arts, to bring criticality and insight through new partnerships. The chapters present a rich anthology of collaborative practices, and creative inquiry. Authors produce accounts of co-designing cultural representations to counter reductive, harmful, and stereotypical images of diversity and heterogeneity. They report achieving this partly through collaborative arts practice such as poetry, theatre, and the visual arts, and also through more familiar approaches such as professional development, and collegial debate. A strong theme throughout is the ability of arts practice and art-informed approaches to build trust, flatten hierarchies, and construct the normativity of linguistic diversity. A recurring consideration is whether the arts, in partnership with applied linguistics and educational studies, can portray the translanguaging realities of people's lives. The collaborative practice such a union demands seeks a discursive space for dialogism and polyphony.

Polyphony is, of course, a concept borrowed from Bakhtin (1984). Bakhtin, a literary critic, showed a clear preference for the polyphonic novel in contrast to other literary forms such as the epic tale. In the former, characters are allowed to speak freely in their own voice in an open-ended process of becoming and development, while in the latter characters are closed and predictable, and under the strict control of the author, who favours his (sic) own voice over those of his characters. In the polyphonic novel there is a 'plurality of independent and unmerged voices and consciousnesses, a genuine polyphony of fully valid voices' (Bakhtin, 1984: 6), while in the epic tale characters are fixed and one-dimensional. For Bakhtin, polyphony introduces multiple voices, and with them, a diversity of speech types.

While Bakhtin was concerned with the genre of the novel, his work is also relevant to the way we author texts in the academy, including the way we represent the lives of those we research. As researchers we face decisions not only about what to say about research participants, but also how to let them say it. One speech type typically excluded in academic discourse is translanguaging. When translanguaging does make an entrance in academic text, it appears as an example of 'data'. Academic articles are not written in a translanguaging style. Indeed they require the removal of the polyphony of voices that have shaped the arguments presented. The academic text prizes the controlling hand of the author, often an individual author who attempts to bring under control the multiple voices that have shaped the research. The academic article has little time for translanguaging as a means to convey its ideas, or to articulate the collaborations typical of the research process.

The authors in this book explore ways to open up the research process so that other voices can be heard in research accounts, not only in terms of content, but also in terms of style. The ability of the arts to put the academy under pressure and explore different forms of representation is a particularly promising direction. This involves finding ways in which the academic voice is backgrounded in favour of the polyphony of everyday life. When we turn to the arts we are provided with different models for the way we represent people, their human relationships, and their translanguaging realities. Nobel prize winner and documentary novelist Svetlana Alexievich (2015) speaks of the polyphony of everyday lives:

> When I walk down the street and catch words, phrases, and exclamations, it fascinates me, and has made me its captive. I love how humans talk. I love the human voice. It is my greatest love and passion. I collect the everyday life of feelings, thoughts, and words. I collect the life of my time.
>
> There are no borders between fact and fabrication, one flows into the other. Witnesses are not impartial. In telling a story, humans create, they wrestle time like a sculptor does marble. They are actors and creators.

As researchers, we are also actors and creators of the stories we choose to tell. Arts practice and arts-informed approaches, collaborations and partnerships promise the production of a polyphony which honours voices from the margins. As García points out in her Foreword, explorations of creative inquiry must go beyond the aesthetic to criticality if they are to result in challenging structural inequalities. If cultural representations are structured by relationships of power and history, ways need to be found to disrupt these relationships in favour of other ways of representing the voices of the excluded and marginalized. Translanguaging is a social practice, an ideological resource, and a political perspective that seeks to interrupt the structures which have closed down the polyphony of everyday voices in favour of the voices of elites, including our own in the academy. If in exploring new relationships, partnerships, and collaborations, new avenues for cultural representation can be found, we have taken the right route. The arts, including literature, poetry, theatre, drama, music, film and visual methods offer new ways to engage with people's complex communicative repertoires and represent the open-endedness of changing lives.

References

Alexievich, S. (2015) *On the Battle Lost*. Nobel Lecture by Svetlana Alexievich, Svenska Akadamien www.nobelprize.org/prizes/literature/2015/alexievich/lecture/ (accessed January 2020).

Bakhtin, M.M. (1984) *Problems of Dostoevsky's Poetics* (C. Emerson, ed. and trans.). Manchester: Manchester University Press.

Index

Page numbers in italic type refer to tables, figures and photographs. Those followed by 'n' refer to notes.